MICHAEL CLAYTON
CUTTING THE COST OF ENERGY
A Practical Guide for the Householder

David & Charles
Newton Abbot   London   North Pomfret (Vt)

British Library Cataloguing in Publication Data

Clayton, Michael
 Cutting the cost of energy.
 1. Dwellings—Great Britain—Heating and
 ventilation
 2. Heating—Cost control
 3 Dwellings—Great Britain—Energy
 conservation
 644'.1   TH7226

ISBN 0-7153-7927-5

© Michael Clayton 1981

All rights reserved. No part of this
publication may be reproduced, stored
in retrieval system, or transmitted,
in any form or by any means, electronic,
mechanical, photocopying, recording or
otherwise, without the prior permission
of David & Charles (Publishers) Limited

Typeset by A.B.M. Typographics Ltd, Hull
and printed in Great Britain
by Redwood Burn Ltd, Trowbridge, Wilts
for David & Charles (Publishers) Limited
Brunel House   Newton Abbot   Devon

Published in the United States of America
by David & Charles Inc
North Pomfret   Vermont 05053   USA

Contents

Introduction		5
1	Insulation	9
2	Solar Heating	26
3	Burning Wood and Other Solid Fuels	63
4	Windpower	72
5	Waterpower	80
6	Methane Fuel	90
7	Heat Pumps	93
8	Central Heating	96
9	Energy in Industry	102
10	Planning Permission	104
Appendix I: Advice on Insulation		108
Appendix II: Solar Energy Companies		110
Index		125

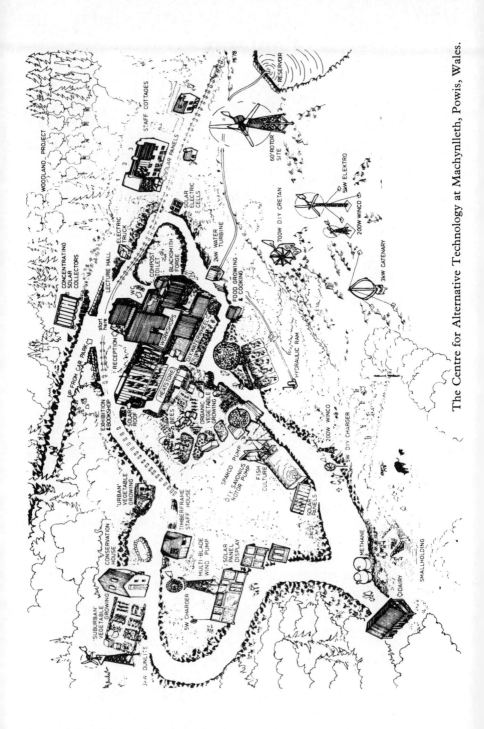

The Centre for Alternative Technology at Machynlleth, Powis, Wales.

Introduction

Energy conservation saves you money and at the same time shows your concern about environment and the future of energy supplies. What you do now may well prompt others in your neighbourhood to follow suit. This book will explain the various approaches to cost-cutting, and it should set you and your family on the path to becoming energy-conscious—to your own and the general benefit.

Imagine a country cottage with water available via two flights of steps downwards from the building. Provision of water would be a real labour. Soon water used in hot water bottles at night would be retained for purposes like washing up. The benefit in terms of water carrying and heating is obvious. But with piped hot water, energy-saving measures are less obvious. If you want to rinse mud off your hands after gardening, you no doubt automatically turn on the hot tap. But you may well have finished before the water runs hot. Yet you have still taken water from the hot tank and that water will remain in the pipeline to go cold—with your water heater making good the heat loss in the tank. The same thing applies when you fill a kettle for any purpose. For making tea there is good reason to start with cold water, yet most people almost certainly take the water from the hot tap. As when rinsing the hands, this probably means that the water into the kettle is cold, but a similar volume of hot water is brought into the pipeline simply to cool off in turn.

Another basic way in which to become energy-conscious is to take note of the 'before and after' of really good lagging of the hot water tank, and as far as possible the hot water pipes. With poor lagging or none at all, heat up the tank to the temperature set on the thermostat of, say, an immersion heater. Then switch off and note how long the water stays adequately warm for most purposes. Almost certainly the water will be cold in 12 to 24 hours. After

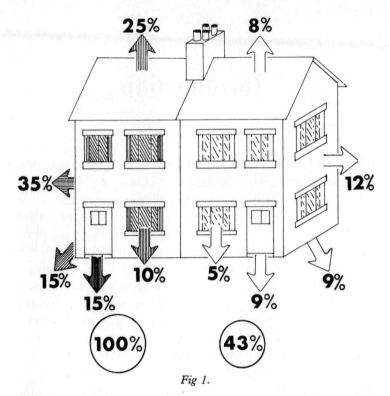

Fig 1.

really good lagging, and again cutting out baths (only), one may find that water that is piping hot on a Monday morning will still be warm enough for washing up, washing oneself and the like right up until the Wednesday—a staggering saving for a modest outlay.

No part of this book should be 'skipped' because, for example, if you have no chimney and therefore think you have no use for solid fuel heating you can be wrong. It is easy these days to install a satisfactory stovepipe with attractive cladding as required, and even have a solid fuel heater freestanding on carpet. Again, this book may show you the advantages—enjoyed in most countries—of using warm air for central heating rather than water, which has the cost of water heating, plumbing, radiators, and a heat exchange system from source-to-water-to-metal-to-air in the rooms—and is slow. In contrast, warm air when not needed can be switched off or adjusted at the touch of a grille slide, and may be turned off at source even for short periods away from the house, as the subsequent redistribution of warm air direct to rooms is virtually immediate.

Fig 1 shows the main sources of heat loss. You will see that double glazing can reduce heat loss by only five per cent. Yet double glazing costs a great deal. The cost benefit would, however, be greater if the windows are facing in the direction of the coldest winds. Another factor is whether the amenity and the appearance of the house are improved by full length, sliding terrace/loggia/patio doors. But the latter must not be confused with energy saving likely to pay for itself within a reasonable period. As one advertiser of cavity wall insulation puts it, the expenditure of £x infilling cavity walls will provide an energy return of more than £3x spent on double glazing. There is also the point that double glazing for heat retention requires the two sheets of glass to be very close, but for soundproofing purposes the sheets should be several inches apart. To have it both ways you need treble glazing, at great cost but with little financial gain in cutting fuel costs.

On the same theme, bear in mind also that heat loss through windows is at its most noticeable on cold winter evenings. Lined curtains or thick drapes will look after this.

As one's outlook begins to take root in the conservation of energy other smaller but significant factors may become apparent. Draught prevention is one, for there is a big difference between draught and controlled ventilation. The economic use of the deep-freeze is another. In the country it has become very much a part of everyday life, permitting purchases of meat at little more than half the butchers' shop prices for each cut, and enabling appropriate garden produce to be preserved. The same applies also to buying in bulk when glut prices exist. But to save the maximum, the deep-freeze should always be kept as full as possible, since then the energy used in keeping things frozen will be at a minimum. Then there is the case of baths versus showers. The latter are much cheaper per person, and although buying an extra energy-consuming device may seem wasteful, it can pay for itself over a period.

But, as you will read, insulation is the first step, the cheapest, and the single most effective saving—and it pays for its modest outlay more quickly than anything else.

1
Insulation

Given the choice between a sieve and a bucket with a slow leak in which to carry liquid, one would opt for the bucket. Call the liquid heating oil for household use and the bucket will still win. But if your house has been built and subsequently maintained without special thought and action about insulation, then it is more like a sieve in the way it squanders heating oil. But suppose we can turn the sieve into a bucket at low cost, and with a slow leak which we will call ventilation, then much less oil will be needed. Heat is expensive and the main, and by far the cheapest, form of saving it is insulation. As energy-saving measures, many forms of insulation have the added attraction for the handyman that they require no help from a craftsman. The implications of insulation need similar study by anyone planning to build from new, for while architects keep abreast of new developments and materials for insulation, the same cannot always be said of the small builder (even though he would probably say the same of the architect).

If a house can be made into a cocoon, then remarkably little space heating is needed, even in winter. Think, for example, of existing sources of warmth: the sun through the windows by day with luck; the heat output from a refrigerator, deepfreeze, or both; the cooker; light bulbs (whose output is nearly all heat rather than light); and so on, according to the individual house. An 'unheated' cocooned house can be quite warm to walk into after a long absence and, most importantly, will warm up very quickly with perhaps one bar of an electric fire, whereas without good insulation double this output would probably be needed all the time to keep just the living room tolerable.

The 'before and after' illustration in the Introduction (Fig 1) gives a reasonably accurate idea of heat loss. The method of making such measurements is quite simple. Each building material has a

'U' factor—a resistance to heat loss (heat conduction). In a typical house like that shown, and allowing for ventilation changing half the air per hour, one should have a power consumption of 11,000 BTU as a maximum. Yet the same basic house could easily be using up some 30,000–40,000 BTU. In many cases, therefore, heating energy may be reduced *to* not *by* 25 per cent.

In a tabular comparison we have these percentages for 'before' and 'after':

Walls	35 to 12
Roof	25 to 8
Floor	15 to 9
Draughts	15 to 9
Windows	10 to 5

At the time of writing, national advertising would incline the uninitiated to put double or treble glazing on a par with cavity wall insulation. Of course, the former you can see and there is the undeniable attraction of, for example, smoothly gliding double glazed french windows. But in terms of cost related to energy factor (or how long it takes to pay for itself), the figures speak for themselves. On this same theme, peak demand for heat in most houses is on winter evenings, when the curtains are drawn across the glazing; if the drapes are of adequate quality or lined, they make a real cut in heat loss through the glass. Two other points should be borne in mind concerning windows. First, that they are the most frequently used ventilators and when opened for ventilation their glazing is of little account. Second, that while the sealed units provide an almost stationary pocket of insulating air, their sound-deadening properties are much less than with a wider gap; but that a wider gap gives poorer heat insulation, since the air can more easily begin a convection movement. This information is not intended to 'knock' double glazing, which should certainly be included in the specifications of a new house (preferably with a third sheet in a noisy situation), but to point out that other insulation techniques are less costly or at least pay for themselves over a shorter period.

Wall insulation

Most sources tell us that the majority of houses have cavity walls. This may be true of modern houses, but the typical brick bond of older houses was solid and, with mortar, the walls a good 13in thick. Walls of older houses of a mixture of stone and rubble were also solid and generally about 2ft thick. Insulation of solid walls presents a major problem in cost, disruption, redecoration and loss of interior space. However, probably fifty per cent of houses are of cavity wall construction. Insulating them is one of the few important jobs you cannot do yourself because of the special equipment required. At the time of writing, there is some dispute about the possible need to air the house well for some days afterwards if the insulating foam contains urea formaldehyde. Adverse effects can vary from skin to breathing trouble. This is something you can check with the contractor.

The Department of Energy (London) has some useful information which, along with information from other sources, has been culled by the author. One point stands out as a safeguard from entering into a contract with a 'cowboy'. Check that the installer is on the Agrément Board list of approved installers, drawn up by the British Standards Institution, for then he will comply with laid down conditions and need only inform the local authority that the work is to be done. Otherwise one must get permission from the local authority before work starts. One cannot help but feel that to use a listed contractor may help to avoid the cowboys, but the selection is yours as are the influences of local circumstances. Always read the small print of the contract especially that relating to long-term guarantees, which are frequently offered.

The benefit of cavity wall insulation will naturally depend on the walls' exposure to weather and the conductivity of the actual wall construction. Here again, a specialist installer will be able to advise. If in any doubt, get an opinion from the local authority surveyor (or his equivalent where you live). In nearly all cases where the structure of the house is involved, it is as well to have a word with the local authority. This is sometimes mandatory, thus explaining the number of newspaper reports of householders being ordered to undo their work.

The example of a friend of the author's living in north-west London illustrates the value of seeking local authority advice first. This friend began to build a small wall on small foundations as the first stage of a quite light glazed conservatory attached to the back of his house. It sounded harmless enough, but the authority got wind of it and told him to stop, with the result that a slanging match ensued. The surveyor then wrote in a conciliatory way and suggested a calmer meeting for discussion. At this he explained his objection. London is built on clay and well built houses have fairly hefty foundations to cope with movement up and down according to season and rainfall. If the lightweight conservatory were to remain attached to the house, it must ride up and down imperceptibly with the house itself. This called for foundations just as wide and deep as those of the main walls. The friend took the point and started again from scratch. Apart from points like this, it is often possible to get the surveyors' department to drop hints about suitable contractors, even if only a short list of those they use themselves for certain tasks.

The value of wall insulation varies according to the distribution of heat in the house. If full central heating is in regular use, then wall insulation will make for real savings *provided* the central heating is adjusted accordingly (unless fully thermostatic). Again, if cavity wall insulation is installed at the same time as a central heating system, then it may well be that a saving can be made on the size and cost of the heating installation.

What is the breakeven point on outlay set against fuel savings? Inevitably, there are several factors to consider. These include the size of the house and nature of outside walls, the effect of weather in the particular location, the costliness of the fuel you use, and how high you like the house temperature to be, especially with central heating. A broad guess based on current prices is five to ten years, but with fuel costs rising so quickly, the breakeven point may tend to become earlier. On the other hand, over a period any building work is also likely to rise in price.

Loft insulation

This is a likely starting point for the handyman if you have a pitched roof, for no great skill is required to cover the loft floor, the cost is comparatively small, and the heat savings can cut a 25 per cent wastage to 8 per cent. Unless it is necessary to use insulating material which has to be machine blown into place, there is no need as a rule to call upon specialist help. But if you prefer not to tackle it yourself, virtually any builder should be able to do it without labour charges reaching the ridge tiles! A flat roof is a much more difficult problem and the cost of effective insulation may outweigh the savings in heating fuel.

Materials for loft insulation consist of mineral or glass fibre rolls of soft, thick matting, or granular material of lightweight plastic such as Vermiculite. Check first on what insulation if any exists in the loft: if there is some, check its thickness; if there is none at all, then before starting work inquire from the local authority about a grant towards the cost. At the time of writing this was £50 in the UK for a private dwelling. Insulation may be as little as 25mm (1in) thick, whereas it should ideally be 100mm (4in), or at least 80mm (3in). The higher the general level of house heating, the greater the thickness of insulation needed. The rolls of material may be obtained easily in these thicknesses.

Before starting work, take some precautions. First, it is better if you can to rig up a decent inspection lamp in a wire cage rather than use a torch. Next, remember to use gloves when handling glass fibre, as this material *is* glass, and although it will not cut in the orthodox sense, it can cause unpleasantly sore hands. When in the loft, tread only on the ceiling joists, not in between (or you may find yourself having to replaster the ceiling of the room below). Measure the distance between the joists and also their length to calculate the material required. Remembering that water pipes and tank(s) will also have to be insulated, measure their dimensions as well. With the pipes you need be concerned only with the size and length of those above the level of the general loft insulation.

If you have settled on granular insulation, measure the depth of the joists. If this is less than the thickness you want for insulation, you have two basic options: either using less insulation than you had planned; or the harder way by putting strips of wood on the joists to raise their height at least between the entrance and main pipe runs and the tank. Boarding on top will then provide safe walkways.

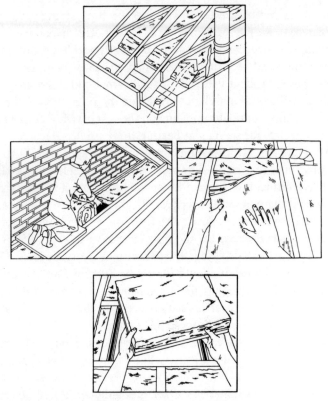

Fig 2. The method of lagging the loft is shown here, but it is wiser to wear gloves. Note that the opening to the loft should also be lagged but fresh air, preferably through the eaves as shown, must be able to circulate round the roof timberwork.

Insulation must not prevent adequate ventilation, for condensation will lead to woodwork damage. Check if there is some daylight at the eaves, for this is air from the exterior, fully effective and not carrying heat with it from the house itself. Alternatively, if you can see the bare slates or tiles, there will be adequate ventilation. But if the slates have been laid over felt, or are boarded, then you must have ventilation from the eaves if you have to bore holes yourself (Fig 2).

The general condition of service essentials in the loft area should be examined at this stage, particularly plumbing and electric wiring. If the main tank is corroded, now is the time for replacement. Quite possibly a metal tank of the same size may be too big to go through the trapdoor, the original having been fitted during the construction of the house. Nowadays rust-free plastic is favoured and may be distorted to get through the trapdoor, or two smaller tanks can be used, linked together. If you are in doubt about wiring, it will be worth a reasonable fee for an electrician or the Electricity Board to take a look.

Tanks and pipes

It used to be almost traditional that cold water pipes and junctions were in or outside the north-east corner of the house and therefore most likely to freeze up. But whatever the layout, insulating the loft will make it much colder. This means that any pipework and water tank(s) must also be protected by insulation. Don't insulate *under* tanks as this may rob them of any remaining warmth from underneath and make them more frost-prone.

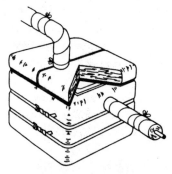

Fig 3. Using mineral or glass fibre material. It is also possible to use mineral or glass fibre quilting secured with wire, tape or netting.

Tanks can be insulated in either of two basic ways, again a choice between granular material and glass fibre or mineral cladding. If the latter is used (Figs 3 and 4), it can be cut to shape and then held in position with tape. The top of the tank must have the same treatment, which first involves a divided lid to clear any expansion pipe entering the tank from the top. The ball valve to the inflow must always be accessible for repair. If a granular insulation is used, the tank must be boxed in (Fig 5). Hardboard or chipboard will do the job but remember that

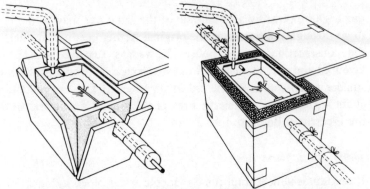

Fig 4. Using sheet insulation material. You can buy pre-cut packs of sheet insulation material to fit the more common sizes of water tank, or you can cut your own casing out of larger pieces. The recommended thickness is 25mm (1in). You should secure the panels with wire, string or tape.
Fig 5. Using loose-fill. If you are using loose-fill, you will need to make a box to contain it. This can be done with hardboard and timber.

granular material will find its way out through the slightest hole. Having taken precautions against such loss, pour the granules slowly, carefully watching for any escape route. It is far from unusual to find that as each bag is poured into the space between the tank and its boxed surround, the granules are silently escaping lower down. Instead of a specially made, probably divided lid covered with insulating material, it is now possible to get floating plastic to stop a layer of ice forming. This may be like a layer of table tennis balls, but your local DIY or builders' merchant will be able to advise.

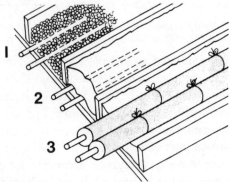

Fig 6. Water pipe insulation: (1) Loose-fill; (2) Fibre; (3) Moulded insulation.

According to the official figures of the pundits, the insulation for tanks should be at least 25mm (1in). This is indeed probably enough to avoid a freeze-up in normal winter weather, but it is wiser to allow for the 'deep-freeze' conditions experienced from time to time and plump for a greater thickness. Water pipe insulation should certainly be thicker—at the least 32mm ($1\frac{1}{4}$in).

There are three basic kinds of pipe insulation—loose-fill, fibre cover and moulded (Fig 6). Fibre cover and loose-fill apply essentially to those horizontal pipes protected by the insulation layer you put down yourself. Moulded insulation is needed if the lay of the pipes is too high to be covered, and to all other free-standing pipes. (Pipe insulation in some houses was included in the original building by boxing in the pipes and using loose-fill—usually sawdust—but today this is unnecessarily expensive.) Exposed pipes can be 'bandaged' with insulation matting, but it is easier to use preformed insulation, which can be clipped or tied to the pipes.

Draughtproofing

While draughts account for about 15 per cent of heat loss, in general they are among the cheapest losses to get rid of. True, with old types of sash window a carpenter may be needed first to readjust the fitting. (At the same time he should be asked to replace sash cords as necessary with nylon cord singed at the ends to avoid fraying; these will then, in effect, last for ever.) The rest of the gaps are easy to seal.

Wastage is mainly through ill-fitting doors, windows, unused open fireplaces, joins between skirting boards and floors, pipe openings in walls and ceilings, and ceiling hatches (usually to the loft).

The first step, obviously, is to hold your hand against the door and window surrounds—and perhaps be surprised at the inrush of cold air. What you will not notice is the reverse, especially high up in doors and windows, of the warm air rushing out. But do not overlook the point that some ventilation is essential, particularly if the heat source is solid fuel, gas, oil, or especially paraffin or bottled gas. Do not block ventilating grilles or air bricks as they are positioned for essential ventilation. Often the latter are provided to allow air to circulate around woodwork.

Condensation is harmful in any circumstances. Bear in mind that oil and bottled gas give off a great deal of moisture, which can lead to condensation. Obvious causes of condensation are bathing, washing and cooking. Condensation can be reduced by opening windows beforehand or installing an extractor fan in kitchen and bathroom. The kitchen extractor can be integrated with a collecting device above the cooker. In the bathroom, a costless system may prove adequate. This is to run some cold water into the bath before turning on the hot tap.

Windows

It has been mentioned that draughtproofing sash windows usually calls for an experienced carpenter or builder. But hinged windows —or those swivelling on a horizontal pivot—are simpler to proof against draught. There are three methods, one using self-adhesive foam strip (Fig 7), and the other two plastic or metal strip (Fig 8). Foam is cheaper and easier, but with constant crushing does not last as long as the alternatives. As windows are opened and closed less frequently than doors, foam is probably the economic answer.

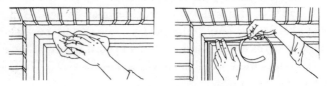

Fig 7. Clean door or window frames before using self-adhesive sealing strip.

Doors

Foam is still the cheapest but needs to be replaced from time to time and the chances are that you will forget to check effectiveness —and therefore replace—often enough. The metal strip is better, and easy to attach when cut to length. Nail holes are provided in the strip and there is a slight crease delineating the part which remains flat on the door surround and that which bends out to seal with the door itself. Remember to leave the lock area clear. When

 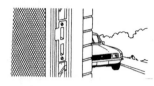

Fig 8. Metal sealing strip has holes for nailing in place. Do not forget to make allowance in advance for the door lock.

the strip is installed, run a screwdriver along the crease to bend out the part which mates with the door as far as necessary. The only likely snag with this system is that you have not removed any lumpy paint from either the door or its surrounds.

A main problem is at the base of exterior doors. One approach is a metal trough, available at virtually all builders' merchants and DIY shops, which is of door thickness but in its flat 'U' shape has a higher lip at the outside. The door base is trimmed to clear the inner lip so that it closes firmly against the higher outer lip. Fastenings are supplied and a DIY man can cut to length and accomplish the job following the instructions. As this is also an excluder of driven rain, the benefit is worthwhile even if a building worker is called in, for the labour would not be expensive. A good answer is a storm porch. This provides a buffer of air—and may also be used for deliveries of mail, parcels, milk and so on. Obviously this is more expensive and it may be impossible in some cases to get the local authority planners' permission. Don't forget the letter box as a source of draught as well as mail. There are also draught excluders that lift up out of the way as the door is opened. The best solution to your particular problems may well be found by a talk both at a builders' merchant and at a DIY shop.

Double glazing

This is one of the main problems concerning insulation, for there are different ways of assessing it. Quoting figures from the Department of Energy, which tie in with those from other sources, a low-priced DIY system for the living room may pay for itself within five to ten years, depending upon the fuel used. But a more expensive proprietary unit installed by the makers is also a matter of décor. To those with their back to the window there is greater comfort, but no more probably than curtains, which would in any case be drawn on cold winter evenings. Double glazed windows can also suffer from condensation between the glass panels even if 'sealed' by the manufacturer, though a DIY system can be moved fairly easily for inter-surface cleaning.

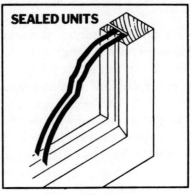

Fig 9.

If a window in any frequently used room needs replacing, it is sensible to check on the price difference of normal single glazing as against a sealed double glazed alternative (Fig 9). But if you suffer from a traffic noise problem then a wider gap is best for sound deadening and so a secondary pane (Fig 10) will be more rewarding. Heat loss benefits may then be less, but as the loss

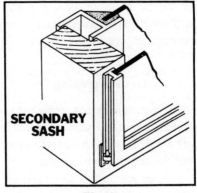

Fig 10.

through windows is small anyway, the sound deadening effect of more widely spaced glass may be your choice. This is where treble glazing comes in—but not for quick returns on energy saving.

Insulating floors

It is estimated that the average house loses about fifteen per cent of its heat through the floor. Often the chief outlets for heat are gaps in the floorboards and skirtings. Sealing with plastic wood or beading is one cure. Another is overall covering with flooring material. With solid floors like concrete or quarry tiles, a good felt or rubber underlay helps greatly—and also makes for an extra touch of luxury when walking over it.

If it is possible to get under a boarded floor, the kind of insulation material mentioned for the loft can be fastened between the joists. But it is important, as in the loft, to permit ventilation round the woodwork. At the least leave air brick vents uncluttered.

Hot water systems

Insulation of cold water tanks and pipes has been mentioned earlier, but from the energy-saving viewpoint, containing heat is of greater importance. It is heat conservation which really triggers one's awareness to the degree of waste hitherto. The starting point is, of course, the hot tank itself, and it is interesting to make a test when the tank has been really well insulated. If the water is brought up to maximum temperature and then used only for washing hands and faces, washing up and so on, the temperature can remain adequate for a couple of days (and a night) on just the one 'heat-up', as mentioned in the Introduction. Further cost savings can be made if you are on the 'White Meter' with its cheaper rates during the night, if the immersion heater is used to heat the tank during those off-peak hours. (It should be mentioned, however, that the White Meter tariff is especially worthwhile if you are up and about using electricity during those hours, or heat water at night, or have night storage heaters.)

An uninsulated hot tank is probably the greatest source of heat loss in many a house. The best lagging will depend on the shape and location of the tank. Many houses, particularly newer bungalows, have a divided tank with a cold water feed at the top with a ball cock under a lid. Usually it is easy to complete their boxing-in using hardboard, but making sure that there are no holes through which granulated packing can escape. When infilled, the box will probably have about four inches of insulation material as a minimum, with more at the squared-up corners.

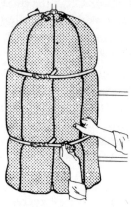

Fig 11.

With sealed tanks there are insulating jackets (Fig 11) to fit most shapes and sizes. A jacket of 2in thickness or less is not adequate but better than nothing if space precludes greater thickness. A really thick jacket will cut heat loss by about 75 per cent (a Department of Energy figure) and pay for itself literally in weeks. If it is possible to box the tank in, the author's preference is nevertheless for inches of granular in-fill. If using a jacket, measure up the tank first and get a 'Kitemark' type showing that it is to British Standards Specification 5615:1978. Most jackets are in segments and therefore slightly fiddly to install, although a collar at the top is helpful. Fitting instructions are supplied but the important points to remember are to smooth the jacket gently into place and to be equally careful not to crush the insulating material tight when doing up the straps supplied. Finally, check that there are no gaps between the segments through which heat can escape. Make sure also that if there is an immersion heater the cap and cables are not covered.

Hot water pipes should be lagged by the method described for cold water pipes, the insulating cover this time keeping heat in rather than frost out.

Controls

Time switches and thermostats (Fig 12) are the basic, important units for heating control. They provide room heating or hot water when you want them, and enable you to use off-peak electricity at the lowest charge if you are on the 'White Meter'. Being easy to adjust, a time control can turn on the heating before you get home even if your comings and goings are irregular, so that you have a warm welcome without heat being wasted while you are out. There are two alternative types of time switch, the more complex one costing about twice that of the simple turn-off-turn-on at preset times.

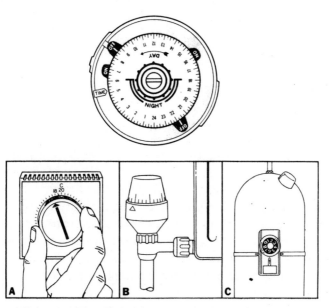

Fig 12. Time switches and controls for setting temperature save money.

The more sophisticated type, called a programmer (Fig 13), can signal a series of 'on' or 'off' periods of the heating system or hot water system or both. As an immersion heater gobbles electricity, a simple time switch costs little in comparison with the savings made, but you must be sure that the switch can handle the heavy electrical load involved. If you cultivate the 'awareness' of heat conservation, you will soon learn to set control switches to be on in time to heat the space adequately by the time you want it, but equally you will set an 'off' in advance of vacating the room.

Argument has existed for years on the subject of the best use of immersion heaters. To leave them on all the time and rely on the built-in thermostat as the sole control? Or to turn them on and off as you think fit? Answer: *don't* leave them on all the time. The hotter the tank water, the greater the heat loss in spite of insulation. Particularly with the overnight off-peak tariff, it obviously pays to have a time switch putting power on at the 'cheap' rate and turning it off before the daytime tariff takes over. Your insulation will keep it hot. For early risers it can also pay to have morning baths (in water heated at the cheap rate) rather than in the evening, when you will probably be paying the full rate. A family can work out a programme compatible with their normal custom.

As opposed to a time switch, a programmer can differentiate between the hot water and the space heating systems. Whatever the make and appearance of the programmer, the functions are similar. There are four positions (Fig 13), for the CH (central heating): a) On 24 hours; b) On once, usually in the morning and off at a suitable time in the evening, usually well before you go to bed so that residual warmth is used up; c) On twice, usually for early morning and then off, followed by another on and off covering late afternoon and evening; d) Off. The HW (hot water) system has the same variety of control but may be at different times in the 24 hours. Usually one can see at a glance what programme is set. With a programmer you can, for example, have one system on all the time and the other for two periods a day. Again, in the summer the heating can be off but the water heating on for two periods as in the winter.

These advantages can cut costs markedly. Early morning hot water is available automatically when you want it: no getting up to switch on or leaving it on all night. The house is warm to return to without heat being on all day. And you save money because you

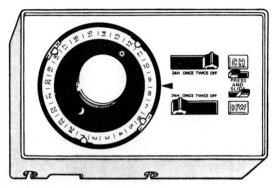

Fig 13. A comprehensive, multi-function heat programmer—best of all.

cannot forget to switch off either system at appropriate times. Experimentation with the settings is a first requirement, and after that making suitable adjustments from season to season.

The Department of Energy has drawn up a wide ranging list of sources of help or advice in the UK. The list does not include manufacturers individually but gives their central organisations, which can provide manufacturers' names. They can also provide broad advice on any particular aspect of insulation and alternative forms of heating. The list is given in Appendix I.

2
Solar Heating

Solar energy harnessed to domestic use has become sufficiently commonplace for many companies to manufacture the ironmongery—and glassware—and to offer control equipment, all complete with an installation service. At the time of writing, more than 70 firms have entered the solar collector business, most since the events of 1973 when energy supplies reached a crisis point, not least for domestic budgets.

Is there enough sun? Half enough, anyway, for domestic hot water, even in temperate latitudes. Variants will include the weather, of course, the efficiency of the system and its size, the quantity of hot water needed, and intelligent use. On the last point, for example, one may choose to have a bath after a reasonable period of sunshine, or wash woollens while the water is only warm, with a hot wash for linen, etc, later.

The choice facing the home owner or occupant is threefold: a commercially produced and installed system; readymade components self-installed; or near-total DIY. The various possibilities will be discussed on the following pages. But it may be said now that should DIY be preferred, there is a further choice between building solar heat collectors comparable with those commercially available or making the system using very cheap secondhand domestic radiators of the right type as a shortcut. However, if the householder is sufficiently practical, the DIY approach may reduce the cost to as little as one third of the commercially installed system.

A southern aspect is needed for solar heat collectors, but this does not mean that some plane of the roof need be south-facing. Solar panels may be wall-mounted (and in this position may even be less intrusive into the design of the house as a whole). Our minds cannot really comprehend the total energy of the sun, but science tells us that even on this planet we get in ten days as much power as remains in all our fossil fuels!

Solar energy is clean. There are no fossil-fumes from a chimney, and on a local—or even world—scale you contribute towards a cleaner atmosphere. This is a side benefit only, now, but likely to be of increasing significance. With national governments taking progressively more interest in promoting research and encouraging energy savings in practical ways such as tax concessions or grants, unnecessary use of fossil fuels may become frowned upon rather in the way that tobacco smoking has acquired an anti-social aura—especially among those who have stopped. For this reason alone, it is as well to contact the local authority on the matter of financial help in any aspect of energy saving.

Fundamental methods

Solar cells

These are perhaps the most efficient way of converting the sun's energy—but also the most expensive. They are usually circular, measuring some 50–75mm (2–3in) in diameter. The technical term for them is photovoltaic. But so many would have to be used —covering the whole roof—that the cost would exceed that of the value of the house by many times over.

Concentrators

The big thought for the future of solar energy is to use mirror-like surfaces of parabolic shape to follow the sun and focus the resultant energy on to a much smaller area. A notable example is in the French Pyrenees, where a temperature of $4,000°C$ is achieved for metal smelting. There are many mundane uses such as boiling water in a black kettle from a simple parabolic reflector. For water heating in quantity, the flow pipe may pass through a trough-shaped reflector. But for the home owner in 'advanced' countries in the northern hemisphere there is a basic snag: the lack of constant sun when required means that the concentrators would need to be mechanised to track the sun.

Solar stills

In the hotter countries salt or brackish water can be distilled quite simply, using solar energy. The still is simply a tank with one side higher than the other, covered with sloping glass. The tank has a black—heat-absorbent—finish. The water inside heats to evaporation point, rises to condense on the glass, and trickles down the glass slope to a collecting 'take-away' channel placed under the glass at the lower side of the tank.

Sunshine cooling

The modern electric or gas refrigerator uses heat to cool the contents. In the same way the power of the sun can be used to cool whole buildings. In its simplest form the system involves an outer south frontage of glass with openings at the top from which the rising warm air can escape. Vents in the base of the wall proper will allow interior air to be sucked out, so that cooler air from more shaded parts of the house—or from north-facing windows—will be drawn into the building. More sophisticated solar air conditioning methods have been developed, but are not yet competitive with conventional refrigeration systems.

Simple solar heating

A basic warming system suitable for the cooler but sunny days of the year can be provided by exploiting the 'greenhouse effect'. Gardeners will have noticed that even on cold days, if there is sunshine, the greenhouse will be warm. This effect can be used if a lean-to greenhouse is built against a south-facing wall to a height to include one or more windows—or french windows (Fig 14). Again, ducts will be needed low in the main wall to allow air to enter the greenhouse area. This time air warms in the greenhouse and passes into the house via the open window(s). An additional benefit is, of course, the greenhouse or conservatory itself.

The basic problem with solar space heating is that most heat is

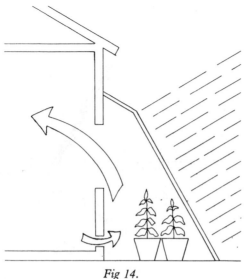

Fig 14.

needed when the sun is not shining. The widely used answer is to store as much heat as possible in water. Whether the water is used only to supply taps or whether it is used in a central heating system, it is still much cheaper to boost lukewarm water to the required temperature than to heat it from cold. The larger the (well insulated) hot water reservoir, the more savings will be made in the long term. When a new house is to be built there is a very sound case for a really big underground storage tank (Fig 15).

The flat plate solar collector

This is the heart of nearly all domestic systems, as well as being especially suitable for swimming pools. On a cost basis flat plate solar collectors pay for themselves most rapidly and their technology has been well proven over the years. The collectors are insulated in weatherproof housings, glazed and mounted to face the sun's position at noon. Once the plumbing and control equipment is complete, they should be long-lasting and virtually maintenance free. Nor do they need tracking mechanisms to follow the path of the sun. Equally attractive is that they lend themselves to competent DIY approaches, either from scratch or by installing readymade units. The flat plate collector is black to absorb radia-

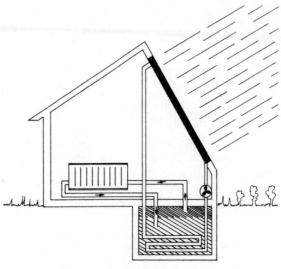

Fig 15.

tion. It is piped with a fluid to take away the energy as heat, and the pipework passes through a secondary hot water tank for storage.

Another feature of solar heat collection favours domestic application because here the temperatures required are not very high: some 45–55°C. The hotter the water gets, the more 'wastage' there will be through subsequent heat loss. Therefore the flat plate solar collectors, operating at *comparatively* low temperatures, are the most efficient.

Some basics

In your schooldays you were told that heat transfer is by three methods only: conduction, convection and radiation. Conduction is by physical contact: to prove it, feel the heat at the handle of a toasting fork, or pick up a saucepan when the handle is of metal and directly connected with the pan. Convection happens with both liquids and gasses (like air), with the warmth rising to the top. In houses radiation is mostly noticed from heat sources of the gas or electric firebar type, and of course from a hot fire in the grate.

Radiant energy is what we get from the sun, and it is this that has to be captured as efficiently as possible. Fortunately, radiant energy knows no barriers and can even travel through a vacuum, which explains why the hot sun heats up the otherwise cold planet on which we live. Sun tan on the human skin demonstrates this. Now we come back to the 'greenhouse effect'. In this the wavelength of radiation is important. The different wavelengths give us the colours of the spectrum, but they also extend into the invisible infra-red, which is too long a wavelength to see, and the ultra-violet, which is too short.

We know that when we see a colour it is because the object we are looking at absorbs the other colours. If the whole spectrum is reflected, we see white, but if it is all absorbed, we see black. Solar heat collectors are painted black so that the maximum amount of heat is absorbed. (In reverse, in the tropics, white is usual for clothing as it reflects the maximum amount of heat and the wearer is much cooler than in a coloured or black covering.) An everyday example of this black versus white can be judged by touching white and black finished metals on a hot sunny day and finding the white much cooler.

Glass and other transparent materials pass solar heat well. But the angle to the sun is important because if the collector is not at right angles to the sun there will be reflection of some of the radiation. In practical terms the collector must be positioned at right angles to the sun at its zenith. In the 'greenhouse effect' heat is trapped because the glass passes a major portion of solar energy, including short wave radiation, but the longer wavelengths are absorbed by whatever may be behind the glass and are therefore not re-radiated. Thus under glass the heat is retained, as in a greenhouse.

We have said that the position of the collectors should be at right angles to the sun at its zenith. But without prohibitively expensive tracking equipment, there must be some compromise. Flat plate collectors absorb warmth even when the sun's rays are partially diffused by the atmosphere. So if one makes the most of sun in high summer then absorption may not be at its peak when most needed. Autumn and spring provide the best aiming points, which in the UK and similar latitudes would require a tilt angle of about 45°. On this basis, any shortfall on the optimum may be countered by using a larger collector area. The same applies if the construction of the house plays a main part in the angle of tilt.

Optimum surface area

The larger the area of the flat plate collectors, the greater the heat catchment, but this is over-simplification and may even lead to problems of cooling in the summer. The rule of thumb method suggests one square foot for each gallon of water to be heated, but to play safe the water may be increased to 1.3 gallons for each square foot (50 to 63 litres per square metre). In normal family terms this means a collector area of three to six square metres. The cost of ancillary equipment remains a virtual constant, so it is wise not to 'economise' on the collector area. However, the greater the mass of water in the collectors, the slower it is to react to the sun's effect. This is important in a temperate climate in which cloud can quickly obscure the sun. In most installations the optimum, setting cost against effect, is an area of four square metres. Remember, though, that this is simply a guide.

Flat plate collectors

A flat plate collector is at the heart of the most commonly used solar heating systems. Its four basic components are: 1) absorption plate; 2) insulation; 3) translucent cover (usually glass); and 4) weatherproof casing.

Working principle

In operation a small amount of solar radiation is lost to the cover, some is lost owing to reflection, but most passes to the absorber plate. The absorbed energy is turned into heat passed from the plate to the water (or other liquid) channels within it or fixed to it. The heated water rises to an outlet pipe, allowing in cooler water from the inlet at the base, thus setting circulation going. Some of the heat absorbed by the plate is lost by the plate's own radiation, etc, but most is turned to good use, depending on the plate's efficiency. The efficiency is measured by the quantity of heat extracted as a fraction of the total solar energy falling on the cover, so if half the total energy is converted the efficiency is 50 per cent. The efficiency of domestic solar panels may vary according to the system from 30 to 50 per cent—a point to raise with manufacturers.

Window glass is hard to better for the covering. There are plastic competitors but for the most part they are shorter lived, lose out on the 'greenhouse effect', and do not like ultra-violet light. Assuming the use of glass, it is better to stick to a single sheet rather than double glazing. The latter may be better when very high temperatures are involved, but such temperatures bring in their train 'overheating' of the inner glass which may then lead to expansion with provision to avoid cracking.

Dirt on the glass must be considered even though rainfall on the sloping surface is a help. Efficiency can fall by 20 per cent if the glass is grubby, so cleaning at reasonably frequent intervals is worthwhile.

There are three main types of flat plate absorber. The basic is an open trough, trickle system; the second is a sandwich of metal with liquid as the 'filling', usually the radiator type using the 'slimline' kind of domestic radiator; and the third is a combination of tubes and sheet. The first type has more problems than advantages; even though it costs little, it is best avoided for the uses suggested in this book. The second is almost certainly the easiest for DIY, and the third is offered by a number of manufacturers of complete collectors but can be home made. In this last type one must remember that aluminium should not be used in contact with copper—the most commonly used metal in many applications for its high conductivity, availability in suitable pipe and strip sizes, and ease of forming into shapes required.

Ideally a solar collector should have a low water capacity for its area, so that the unit as a whole will react quickly to improved radiation from the sun. The single panel 'slimline' radiator (*not* the double panel type or the oldstyle segmented radiator) holds more than the ideal of 2.5 litres per square metre (0.4 pints per square foot), but for all that some similar solar collectors on the market have an even greater liquid capacity. Some of the terms bandied about among the glut of quite new entrants into the collector manufacturing field include 'selective surfaces' and 'high efficiency'. Mostly such labels relate to absorbers expected to cope with the very high temperatures required in industry and other special applications. For example, the standard 'selective surface' in domestic terms is a good matt black paint rather than a flashing of electroplate to cut down loss by reducing radiation emission from the collector.

Long life

Economics demand that the solar heat system last much longer than needed to cover the outlay by energy cost savings. Time was when one could reasonably accurately calculate how many years would be needed to cover the outlay but now, with energy prices rising so quickly, a forecast of five years may well turn out to be two years too many. But this is no reason not to plan for longevity and to cut important corners. Faults may develop inside or outside the collectors, such as weathering of the case and the glazing joints, the corrosive effect of the fluid, electro-chemical problems from mixing different metals in the system (such as water flow from copper to aluminium), the combined effect of air and water, and so on. For the DIY man probably the two most important points are to avoid aluminium and to ensure that the case—and the bonding of the glass to it—is faultless.

Circulation

The water heated in the collector must obviously be circulated, either to the point of usage or to a storage tank. There are two methods, the first to leave it to nature for hot water to rise and let in the falling cold water underneath, and the second to use a little pump. Although a pump and its controls add to the total cost, much of the extra can be offset against the smaller bore piping which can be used with forced circulation. A pump also improves efficiency, for it will cut in even in a short spell of warmer water in the collector; this water could easily have lost its extra heat if left to the slower thermosyphon or natural circulation. Forced circulation also simplifies installation of other components which, in a thermosyphon system, would have to be one higher than another in heat terms for the circulation to operate. The thermosyphon system calls for low mounted collectors with a high enough loft to house two hot tanks significantly higher and two further tanks (one an expansion unit and the other the cold water feed). On balance, therefore, the answer is forced circulation by a pump.

The pump

Orthodox central heating systems require a pump for circulation so there is no problem of availability or choice. Operation is electric and the pump's power rating is based on the resistance to flow measured as the 'head' of water to be raised. With a system having the storage tank in the attic close to the collectors a suitable pump would be rated 3.5 metres. The most common pipe size in a forced system is 15mm ($\frac{1}{2}$in) and so, if pipe runs between collectors and the store are long, a pump rated at 4.5 metres would do the job. The use of microbore pipe raises resistance to flow and a pump with a head rating of 5.5 metres would be needed. A useful alternative is the variable head pump. Many of the larger manufacturers can now offer these at prices similar to fixed head types.

Finally, it is also possible to get a pump with a two-way flow speed selector. With this, settings on installation should be lowest speed and head rating. If too low, the temperature between collector inlet and outlet will be high, and the speed should be increased. Beyond this, the rating should be increased.

Pump control

An off or on control is needed so that circulation occurs only when the outlet temperature from the collectors is appreciably higher than in the storage tank. In a country in which the temperature pattern is predictable it would be possible to switch the pump on in the morning and off at dusk. But in temperate climates daytime radiation varies and there is therefore the real danger that in cloudy conditions—with the pump running—the conserved heat will be dissipated. The best solution to this control problem is a unit which checks the difference between the temperature in the storage tank and that in the outlet of the collectors. This 'black box' controller is 'fed' from mains electricity, with connections to the pump and also to sensors to compare temperatures between collector output and storage tank. The pump is then out of action unless the temperature at the collector outlet is sufficiently higher than in the storage tank.

The next task is to determine what the temperature difference should be between the two. Normally this is between 3 and 5°C.

The differential must more than allow for the electrical cost of running the pump and, bearing in mind any heat loss from the piping en route, there must be a genuine heat gain in the storage tank. A further aim is to avoid 'hunting' by the pump (switching on and off without practical gain). Some pumps avoid hunting by having a time delay incorporated. With this type of pump the delay is adjusted to take into account how long a period the flow takes to travel from the collector to the tank inlet. This can be measured by stop watch, switching the pump off on the night before and then, on a really sunny morning and when the liquid in the collector is hot, switching the pump on by hand and checking the time taken for the inlet to the storage tank to start getting warm to the touch.

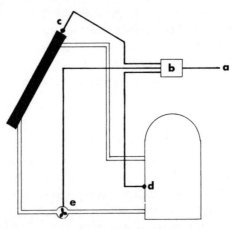

Fig 16. Pump control unit: (a) Mains electricity; (b) Controller; (c) Sensor on solar collector; (d) Sensor on tank; (e) Pump.

The sensors for temperature differential must be fixed not only securely but with good thermal efficiency. There is a choice of methods. The best sensors would intrude into the water itself, but this degree of finesse and complication is not necessary in a straightforward domestic installation. They may be strapped on firmly, or a copper tube may be soldered on at each point with the sensor fastened into its tube with an epoxy resin. The sensor at the storage tank should be attached halfway between the tank's inlet and outlet. At the collector it may be possible to effect the connection just inside the casing. Failing that, it must be as close to the hot outlet as possible.

Frost precautions

In terms of antifreeze protection, it is best to regard a solar collector in the same way as one would a car radiator. On clear, cold winter nights, trouble from freezing may be just as damaging. There are alternative methods of frost protection, such as insulated covers for the collectors, but as a closed circuit is in use, the obvious answer is a solution of water and antifreeze compatible with the materials with which it is in contact. The most commonly used antifreeze is of the ethylene-glycol type. Do *not* use the cheaper type based on methanol (because of danger which need not be explained here). With ethylene-glycol antifreeze the only danger is in the event of a leak from the heat exchanger in the storage tank into the water. Even then the danger applies only if the water is drunk in considerable quantities—and who drinks in such quantities from a hot tap? True, the supply may be used for hot beverages but malfunction of the solar heating performance would show up the fault quickly enough to induce caution. The normal suggested proportion of ethylene-glycol in the water lies between 20 and 25 per cent, which may be interpreted as up to 9 litres (2 gallons) in the average domestic system capacity of about 36 litres (8 gallons).

Storing the heat

Of the two items of prime importance in a solar heating system, one is collection and the other heat storage. The selection of the storage tank—its size and type—must therefore be made with care. Basically, one uses the sun's radiation when it is available to heat the largest practical weight of solid or liquid material and then withdraws heat from this source at night and when the sun is obscured by heavy cloud or fog. An almost exact parallel is the electrical night storage heater, which is warmed at a lower cost rate during off-peak hours and continues to give up its heat for hours afterwards. When solid material is used—like the off-peak storage heater—there is virtually no way of controlling the heat output except when a fan is incorporated to speed up heat output when required. When less heat is needed, say if the weather suddenly turns warm again, it may even be necessary to waste the heat by dissipating it through open windows.

For this reason alone it is better to store the heat in water in a well insulated tank. This method has the further advantage that water absorbs more energy per degree rise in temperature than, say, the firebricks of a solid store. One of the more famous storage systems was achieved by an energy-conscious citizen of New Mexico. His south-facing wall incorporates a filling of oil drums containing water. In the considerable heat of the day they warm up and also help insulate the house interior from the sun. In that area, however, the nights are very cold by comparison so at nightfall a shutter is brought down outside the drums. Radiant heat from the drums keeps the house warm at night. Interesting and so simple—but not the answer in a temperate climate.

There is a lesson to be learned from the exceptional suitability of solar heating for swimming pools. In this case the mass of water is so great that warmth will be retained for long periods when the solar heat source is not effective. Efficiency can be further increased by the use of floating insulation when the pool is not in use or—as at night—not being heated. In domestic heating open tanks may also be insulated on the surface of the water, of which more presently. Experiments are being made with under-floor tanks, suitable for inclusion in new houses or those with a now old-fashioned cellar. There are snags in properly insulating this heat on a warm day from the occupants above, and the cost is clearly high.

For all that, a comparable system has been used to good effect in greenhouses. This involves digging down and then infilling with a suitable size of graded stone, through which air must be able to pass freely. The paving over it stops short of the sides of the greenhouse and, in the centre, allows an opening into a vertical mesh cylinder equipped under floor with a simple fan. Refinements may be included but basically daytime warmth from the 'greenhouse effect' is circulated by fan through the underfloor stone. At nightfall the fan is reversed and the warmth from the stone comes up through the unpaved edges of the greenhouse. For much of the year there is no need to use expensive fuel for heat, particularly if the avoidance of frost is the prime concern.

But for a typical dwelling the only practical heat store is water. For this a copper cylinder, although expensive, is the most satisfactory answer—especially if the pipework is also copper. A galvanised iron tank will tend to corrode in time when used in conjunction with copper piping. Other materials such as glass fibre or stainless steel are, at the time of writing, no cheaper than copper. Size is another important factor. One method is to relate the storage volume to the surface area of the collectors. Thus one should allow 50 to 60 litres per square metre (1gal to 1.3gal per square foot). The other approach is the quantity of warm or hot water required per person. This may vary enormously from person to person according to bathing frequency, which is related to the clean or dirty nature of their work—and play—and must really be taken into account when deciding the area needed for collector surfaces.

With size comes weight, as water is heavy stuff. Metric measure is as always convenient, for each litre weighs 1kg. A cubic foot weighs about 10lb. A typical domestic hot tank holds 200 litres weighing 200kg—in Imperial measure nearly 4cwt. The storage tank must therefore be mounted so that its supports are not overloaded. Probably it can be placed over structually strong dividing walls. In any event the weight should be spread over a number of rafters by the simple use of bearers on edge. Bearers are not enough for plastic tanks; the whole base should be supported on something strong, such as chipboard.

Heat exchanger

To understand this term bear in mind that the best known is an ordinary car radiator. Another is a central heating radiator. Both cool the hot water inside by dispersing it to the surrounding air. A comparable unit in a refrigerator does the same, albeit using heat with the reverse effect. So we have a heat exchanger in a solar collector and also in the storage tank to which it is connected. The heat exchanger simply transfers heat from one medium to another. In a normal domestic installation the water and antifreeze mixture is heated by the heat exchanger in the collector and discharges its warmth into the heat exchanger in the storage tank.

The origin of the technique dates back to an almost accidental discovery by Faraday, even though it was a hundred years later that it was put into practice as a refrigerator. Faraday used a test tube shaped like an inverted 'U', one leg containing silver chloride saturated with ammonia which, when heated, released the ammonia as gas into the other leg which was water-cooled. The ammonia then became liquid. Faraday switched off the heat and later found that the liquid ammonia had gone and in the process had caused ice to be formed on that end of the test tube. The ammonia had gone back to the silver chloride taking heat with it and freezing the water. This was the first absorption refrigeration system. Later it was found that water could take the place of silver chloride, and modern refrigeration was born.

Nowadays experiments are being made to use solar power in cooling systems during heatwaves, in other words to provide solar-powered air conditioning systems. In two of the hottest areas of Australia, water-saturated rocks are used, in each case in conjunction with two electric fans. A suitable application of the principle has yet to be developed for temperate climates.

As the heat exchangers in the collector panels are in a closed circuit with the solar hot water tank, there must be a heat exchanger also in the storage tank to extract the heat. The most common type is a coil of copper tubing in the lower half of the tank. It is not possible for all the heat from the collectors to be given up to the water in the storage tank, but the figure may be as high as 90 per cent if the material used is copper and the rate of flow from the collectors is high. The basic factors affecting the heat exchanger performance are the temperature difference between that in the incoming water and that in the tank; the conductivity of the coil (copper best); surface area of the exchanger (the greater the better); and the rate of flow just mentioned.

Probably the tank chosen for storage will be one already fitted with a copper coil heat exchanger, but if a reasonably competent DIY man picks up a plain tank cheaply, he will need to know where to place the heat exchanger. As with an electric immersion heater, it should be low down in the coolest part of the tank. With a solar heated supply this is especially important because the incoming 'hot' water may be only a few degrees warmer than that in the tank, so the most efficient heat exchange will be to the coldest water.

Selecting a system

There are various advocates, various arguments for the different ways in which solar heating may be incorporated into existing hot water systems. Among the factors are cost, of course; the type of the existing hot water source; frost precautions (which can be complex in themselves); and the height relationship between the existing hot water tank and possible locations for solar panels. Although the most expensive, almost certainly the best is to incorporate a secondary hot tank for the solar heated water and to feed this in to the main hot tank. In this way the hot tank is fed from the solar heated water so that in ideal conditions there is no need for a thermostatic back-up such as an immersion heater to be called into use. If it is, then it has only to raise the temperature of warm or lukewarm water rather than the supply from the standard cold water tank.

If the main hot tank is fed directly from the collector panels it means that the immersion heater, or other hot supply, must be remounted near the top of the tank while the heating element from the collectors is installed near the base. Were this not done, the warm water from the collectors would naturally rise to the top and the immersion heater at the base would continue to be 'on' in the cold layer of water. This method has the further problem: in weather that fails to produce a significant amount of solar heated water the total hot water available will be only that in the upper part of the tank, heated by the primary source (eg immersion heater).

Cost may be cut considerably by using a direct, thermosyphon system but this means that the solar panels must be well below the level of the existing hot tank—which is rarely feasible. Normally the hot tank is at bathroom level, hence below the lowest part of the roof, which is the likely mounting area for the collectors. This system also means that the same water is circulating throughout the system, causing frost problems. The simplest way round this problem is to use a closed circuit through the collectors, thus

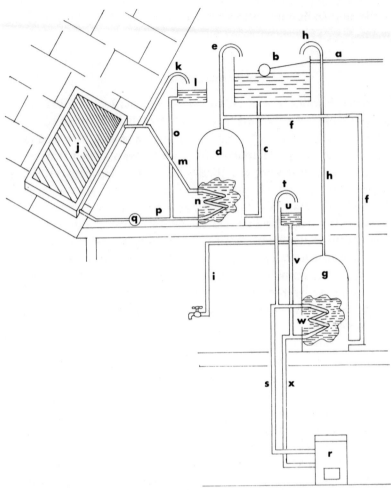

Fig 17. Two tank system: (a) Mains supply; (b) Cold feed tank; (c) Cold feed to solar tank; (d) Solar storage tank; (e) Solar tank vent; (f) Pre-heated water to hot tank; (g) Hot tank; (h) Hot tank vent; (i) Hot water to taps; (j) Solar collectors; (k) Pre-heat system vent; (l) Expansion tank; (m) Flow from collectors; (n) Solar heat exchanger; (o) Pre-heat system feed; (p) Return to collectors; (q) Pump; (r) Boiler; (s) Flow from boiler; (t) Boiler system vent; (u) Expansion tank; (v) Boiler system feed; (w) Heat exchanger; (x) Return to boiler.

permitting the use of antifreeze, and feeding this through a coil heat exchanger low in the hot tank. But here we are back to the other problem of having the primary or backup heat exchanger having to be mounted in the upper part of the storage tank. This means a good deal of plumbing alteration and, with a closed circuit, the installation of an open-topped expansion tank above the system.

Another approach to direct systems is to pipe solar heated water to extra hot taps alongside existing ones, but corrosion will be intensified by air which is always entrained in the water flow. Again, the cost may be put to better use in a more sophisticated system. (With a closed circuit corrosion is limited to the ill-effect only of the air entrained at the actual time of filling the closed circuit.)

From just about every point of view, then, earnest consideration must be given to the more expensive but much more effective and satisfactory two-tank system, with a separate tank for solar heated water. The coldest water from the standard hot tank is then fed to the bottom of the solar tank. This in turn feeds back the warmest water from the top back to the lowest part of the normal hot tank. This means the cost of another tank, a pump, temperature sensors, and a control to 'tell' the pump when to start and stop.

Location of solar collectors

Solar panels are not only the most expensive items, but their location and mounting are fundamental to the success of any solar heating system. The panels must face south, and be as near to their own hot tank as possible. While it is certainly possible to mount the panels even at ground level, the roof is preferable for obvious reasons. The south facing roof is also less likely to be put in the shade for long periods by overhanging trees or other obstructions. As the panels should face south, or certainly not more than 30 deg out of true, the DIY man or builder may have to 'angle' them.

Getting them on to the roof is a major task, for on average they weigh some 70lb each complete. Provision of clips with which to hold them in place will be dealt with later but, assuming that clips are in place for the lower edge of each panel, the lifting operation is the main problem—for panels are far too costly to risk dropping one. The cheapest method, although the most hazardous, is to use

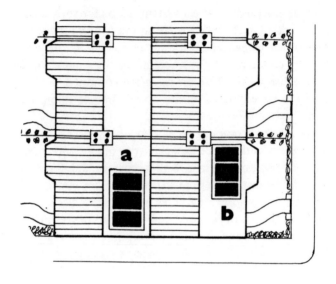

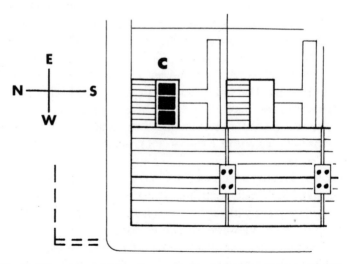

Fig 18. Plan views of (a) gulley roof; (b) pitched roof; (c) outbuilding.

two strong sets of ladders and also a pair of roof ladders. Two men with rope can do the rest. A popular alternative is to use scaffolding. This can be assembled by the suppliers as a column with wheels so that it can be moved to put panels side by side. It is important to secure the wheels at each lifting point. But scaffolding is expensive to hire, have assembled and then taken to pieces again, and is usually hired by the week, whereas half a day is plenty.

The third choice is the most attractive if circumstances permit. This is to use a hydraulic platform of some sort. In the country, farm houses usually have a low eaves level, and the bucket attachment on the front of a tractor provides not only the lifting power but also room for a panel, two men and their tools. In urban conditions one can call local plant hire companies for a quote for a suitable hydraulic lifting device (according to the nearness of access for the machine to the south facing wall) for a period of only an hour or so.

Obviously you will site the panels not only with regard to their aspect to the sun but also with regard to their appearance. Usually, if they blend with the shape of the roof, wall or outhouse, there will be few problems. While most collector panels are mounted over the tiles or slates of the roof, if it seems desirable slates may be removed so that the panels are mounted directly on to the rafters, with the necessary flashing of lead or zinc (both preferable to self-adhesive modern alternatives). Surface mounting is quicker and cheaper. For this there are metal clips available which can be fastened beforehand to the timberwork of the roof that will be supporting the lower and upper parts of the panels. If these clips are in place, at least temporary mounting of the panels—say by hydraulic means—becomes quick and easy. But always beware of the wind, particularly on the lee side; its lifting force can be unexpectedly high, with possibly catastrophic results to the panel being handled.

On a flat roof the panels must be propped up at the optimum angle. The chances are that the panel supplier can provide suitable metal frames but, particularly if you make your own panels, you will be likely to make wooden frames for the mounting angle. In this event remember not only to use wood preservative but to mount the wood on non-corrosive material to avoid unventilated wood being in permanent contact with the roofing material. Pads can be made of lead, nylon, roofing felt, or whatever you think compatible with the roof surface and the wood of the frames.

Alternatively, the nature of the building, its site and perhaps existing plumbing may call for wall mounting. This in turn will also require a mounting structure to angle the panels correctly. If wall-mounted, the panels are likely to be more conspicuous and some thought should be given to effecting an acceptable blend of the panels and the house as a whole (Fig 19). If possible, line up

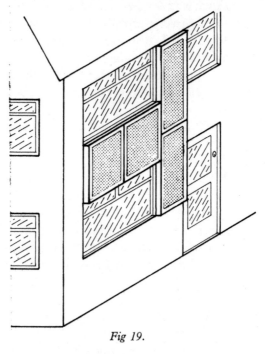

Fig 19.

the panels directly across the upper edge of windows, or perhaps taking up the space between downstairs and upstairs windows. If the wall has a door and perhaps a loggia at its base, it may well be possible to mount the panels as a shelter for the door (Fig 20) or even as a partial sunshade for the loggia (terrace or patio). It is tempting to mount collectors flat against a wall. They could well be effective in high summer, but a tilt of, say, 45 deg will about double their usefulness during the year as a whole. Put another

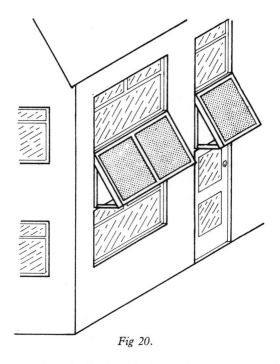

Fig 20.

way, if mounted vertically double the collector area should really be used. Of course, wall mounting is likely when only a gable-end (often called pine-end) is facing south and vertical mounting might be necessary if the panels would otherwise jut out into a right of way, for example. Then one would have to do one's sums again to decide for or against a solar system.

DIY collectors

A basic consideration of DIY is just how much you want to DIY! If skills or time are limited, you will find in the list of manufacturers given in Appendix II (p.110) companies who can provide parts ready for simplifying the task. At the other end of the scale, for the quite skilled enthusiast with the basic ability, facilities and initiative to build panels virtually from scratch, the cheapest basic component is a secondhand 'slimline' domestic radiator for each panel. These may be obtained for a nominal price as scrap. The old fashioned segmented type of radiator is no use, nor is the double-panelled sort. And the single 'slimline' radiator should be of the type holding the minimum amount of water. This type will react most quickly to short periods of sunshine.

When you have a suitable radiator, the next task is to provide the right inlet and exit junctions in the right places and to blank off unwanted holes, such as the little one originally intended as an air bleeder in the central heating system. A secondhand radiator is likely to have connections in inch rather than metric bore sizes, which would ideally be ¾in or 1in in each corner. Metal plugs, suitably threaded, can be used to block off unwanted holes. Often the radiator available may have, in addition to the bleed vent, only two holes—both in line at what was the bottom of the radiator. Use of these raises complications, including a pumped high flow system. It is better to block one and drill a fresh one diagonally opposite the remaining hole, so that coolest water enters at the base and the warmed water leaves from the top, either to the solar tank or to supply the next panel. Brazing 150mm (6in) copper pipes to extend inlet and outlet is the next stage, and if this is beyond your skill or facilities, a local workshop will no doubt do it for you. (Garages and workshops specialising in crash damage are likely places to ask.)

Points to watch

The size of converted radiators will obviously affect the number needed for the required total panel area. Given that the combination of panels will fit the space available, it seems that the larger the radiators the better because the bigger the fewer. But there are pitfalls. One is the overall weight of the finished panel, having in mind the method of lifting and the final site. A flat roof is more suitable for large, heavy single units than a pitched roof. As a rule of thumb, $10ft^2$ is about the reasonable limit for handling in awkward places. But larger sizes effect their own economies. These may handsomely offset the extra cost of some items, like glass, which may be cheaper in smaller, standard sizes. Horticultural glass costs half as much as window glass and comes in standard sizes. Your merchant can give you the standard 'Dutch Light' sizes so that you can balance one cost against another.

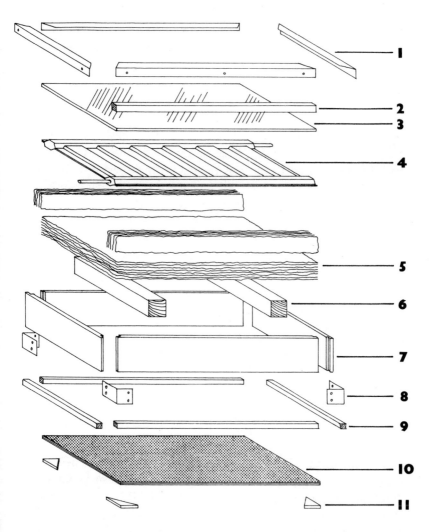

Fig 21. Components of a DIY collector: (1) Glazing angle; (2) Softwood strip; (3) Glass; (4) Absorber panel; (5) Insulation; (6) Absorber support; (7) Side panels; (8) Corner angles; (9) Seating for base; (10) Base; (11) Bracing for corners.

Construction

Primary needs for a DIY complete panel are a baseplate, wooden frame and glazed top. Inside is a layer of insulation material below the matt black painted radiator (now collector or absorber) panel.

Incidentals are also illustrated (Fig 21), including wood supports for the absorber, angle surround for the glazing, a spacer to hold the glass off the absorber, and corner angles and bracing. Only wood has so far been mentioned as prime material for the case whereas glass fibre or aluminium are the materials mainly used for factory-made panels. Should you have had practice with either of these materials then well and good, but most people will be more at home working with easily obtainable wood. Wood will, however, need to be well treated with preservative before painting and thereafter included whenever exterior paintwork is refurbished.

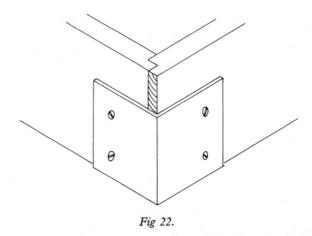

Fig 22.

The choice of woods for the case needs some care. The sides may be made of softwood or resin-bonded chipboard, but the extra cost of hardwood is probably justified. (Bear in mind that most timber merchants deal only in softwoods, but they can put you in touch with the nearest source of hardwoods or order on your behalf.) Work to the internal measurements, whatever the materials used, and begin with the baseboard carefully 'squared up'. The DIY man will know how to do this—and if the reader does not, then he may be taking on a little too much! However, one method is to mark out the shape and then check that the diagonals match up. Teesquare and setsquare are desirable, but one can fall back on Pythagoras for the right angles, remembering that the simplest trio of dimensions is three, four, five.

The base will be protected if it is within the side panels, so the latter should be sawn to fit snugly around it. The side panels can be attached to each other (Fig 22) by simple lap joints, cut at accurate right angles with a tenon saw. The joints should be glued in accordance with the glue manufacturer's instructions and then nailed. Triangles of plywood (shown in Fig 21) can be used for bracing the base of each corner. At this stage something is needed on which the absorber supports can rest, if only for accurate location. The absorber supports are secured by screwing through the side panels. They should be at least 75mm (3in) from each end to leave room for getting at the absorber connections. Also allow for the thickness of the absorber plus an air gap of not less than 13mm ($\frac{1}{2}$) between the upper face of the absorber and the glass covering.

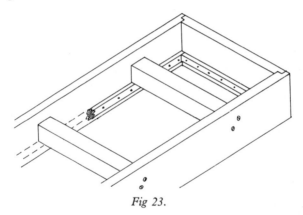

Fig 23.

Seating strips for the baseboard should now be nailed on, with their lower edge some 75mm (3in) below the top of the absorber support strips to allow room for the insulation material (Figs 21 and 23). The baseboard can be screwed on after an application of sealing mastic to ensure impregnability from the weather. Next comes the all-important layer of insulation, which should be 75mm (3in) of glass fibre or mineral wool turned up at the edges for total insulation. You may have enough of this material left over from insulating the roof of the house. In any event, remember that an invisibly fine dust can be given off which can penetrate the hands and also reach the lungs. Therefore wear gloves and some suitable protection over the mouth and nose, such as thickish gauze held with string or elastic.

The absorber can now be positioned on its supports, and the simplest method is to tap in nails above and below as necessary. Then mark precisely on the side panels the positions of the inlet and outlet and drill appropriate holes in the casing. Once this is done aluminium angles should be screwed in place for extra strength and protection of the joints. All this preparatory work helps in the long term to give not only rigidity but weather protection. For this reason the joints should also be covered with mastic on the inside. The aluminium angles should be flush with the base of the frame and at the top should allow only for the glazing angle pieces, which are the last items to be fixed (Fig 25). The angles will, of course, have to be drilled first as necessary for the inlet and outlet pipes.

Particularly with secondhand radiator panels for the absorbers, special care must be taken to ensure that inlet and outlet really are watertight. In the final job compression fittings will be used, with PTFE tape as an extra sealant. Assuming four holes, the other two will be plugged with appropriately threaded bolts. To check on leaks, close one compression fitting with a blanking disc, then fill the radiator through the remaining opening. Stand the unit in a dry place and on a dry surface for as much as an hour or so. Then, if there is no leak, blank off the other outlet, with the unit now the other way up, and test again. Tedious but very important.

At this stage the panel(s) can be thoroughly cleaned for painting. Remove grease, and wire brush rust and loose paintwork. Then apply a primer, followed by thorough painting in matt black on the upper surface facing the sun. Ordinary blackboard paint is as good as anything for the final coating always assuming that the system never gets dry and therefore exceptionally hot. If there is any likelihood of this, heat resistant paints such as those for black-painted stoves can be used.

When the upper side (only) of the absorber has been painted, it can be put in the casing on its supports and well snuggled in insulation. Copper tube of 22mm should now be pushed through the holes already drilled in the casing, and attached by the compression fittings to make leak-proof joints. The length of the copper

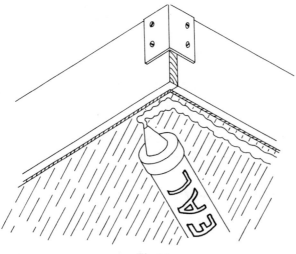

Fig 24.

piping protruding from the casing will depend upon your ultimate plumbing design, but should not be less than 50mm (2in) from the outside of the casing.

Glazing can now be installed. Avoid ordinary putty because its life in this function is not long enough to justify moderately lower cost. Rather use a silicone sealant, which is much longer lasting. If covered by the metal angles helping to secure the glass it will

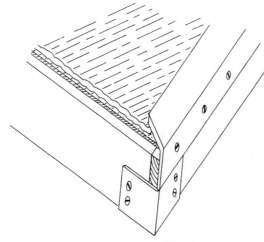

Fig 25.

last even longer. These metal glazing angles are of aluminium, chamfered at the corners when cut to size. Before glazing, nail the wooden support spacer in place. Make sure it is true with the top of the main casing, so that when the aluminium angle is fastened, there will be no irregularity to cause the glass to crack. With the glass in place extrude a thin line of sealant near the edge but not touching the wood (Fig 24). The pre-drilled aluminium glazing angle pieces can now be screwed into place to complete the whole job (Fig 25).

Note on secondhand radiators

A central heating radiator, just like a car radiator, can become scaled internally during use and, unless treated, act inefficiently in its role as a heat exchanger. It is commonsense therefore to descale before starting work to install a secondhand radiator as the collector panel in a solar heated system. In a hard water area the scaling may be like that in a kettle. However, proprietary preparations can be used, following the manufacturers' directions. An alternative method requires dilute acid such as 10 per cent hydrochloric acid. The radiator is filled with this quite weak but safe mixture and left for half a day or so. Then it is emptied and flushed out with clean water before being refilled with plain water. From this time until the radiator is incorporated in the complete solar collector it should be left full of water. Otherwise the acid-cleaned surfaces will rust in air more quickly than untreated surfaces. If using the acid method remember to avoid the mixture contacting clothes or skin (particularly near the eyes). Avoid slopping the mixture and in any case wash immediately the job is finished.

DIY collectors in copper

Copper has a big attraction as primary material for a heat exchanger mainly because heat transfer—or conductivity—is particularly high. In addition, relevant plumbing is likely to be of the same material, thus obviating any adverse chemical reactions between different metals. Both piping and the necessary copper strips for a collector will be bought new and therefore long, trouble-free life may be expected if the construction is sound in the first place. Greater efficiency will also be obtained.

There are two basic ways of using copper. The first uses a horizontal pipe at top and bottom, linked by straight copper pipes of smaller diameter, each of which requires a Tee-joint to connect with the two horizontal pipes (Fig 26). The alternative involves the

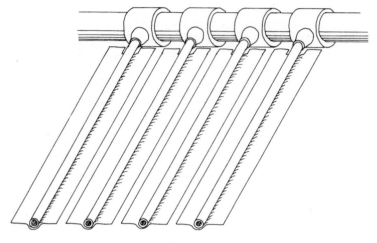

Fig 26.

simpler task of making a zigzag (or serpentine) tube, with fewer joints and therefore fewer problems. On a cost basis the use of copper in this way is about the same as using a new domestic-type radiator panel. Although it means more work, for the real DIY man the extra work is well worthwhile.

The serpentine tube

When bent into its serpentine shape the copper piping is bonded to copper sheet or strip so that radiation on the sheet will quickly pass to the pipe and therefore to the water in it; this makes for maximum solar energy collection. If the serpentine tube is not formed in a flat plane there will be problems in bonding to the copper sheet, which must itself be kept flat. The more bends there are in each section of copper pipe, the more difficult it is to keep the final effect flat.

To avoid these problems, it may be helpful to make use of the standard length in which copper pipe is supplied. For a size of final casing similar to that used for a domestic type radiator conversion, two standard length 3m (nearly 10ft) 15mm copper pipes can be used. Each tube will require only one bend and the two will be connected by a pair of elbow fittings, a total task within the capability of DIY (Fig 27).

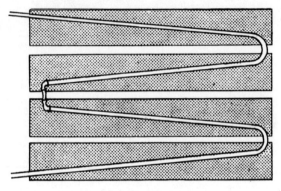

Fig 27.

The elbow fittings will require a connecting piece of pipe, so first saw off 125mm (5in) from one pipe and keep it handy. Then measure three points on each pipe, making a mark at 1,260mm (4ft 1½in), 1,360mm (4ft 5½in) and 1,460mm (4ft 9½in). These measurements mark the middle and the ends of each slightly splayed U-bend. The sort of bender used by a plumber is needed to form the bends. No doubt if necessary a plumber will do this for you. To get the right spread the marks for the start and finish of each bend should be about 125mm (5in) apart while the ends of each tube should be about 175mm (7in) apart. Each splayed U-bend provides for warm water rising. Even though a pump is used in the system, the fallaway provided by the 'snaky' shape described makes for easy draining, should this ever be necessary, and for smooth water flow (Fig 28). Each bend is located so that one leg is longer than the other—in one case 125mm (5in) which was cut to join the elbows—but this is to allow for a longer section for the inlet and outlet. The two shorter sides of the pipes are side by side, joined by the elbows and the 125mm (5in) pipe off-cut. A blow-

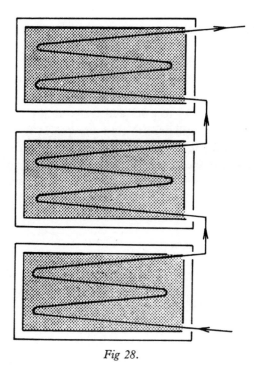

Fig 28.

torch is used to solder the elbows in place and, here again, you may need some help to ensure that they are really watertight and will last indefinitely. The pair of U-bends is encased within the frame of the solar panel as a whole.

Bonding waterways to copper sheet collector

Copper is the natural metal to which to attach the piping. It is easy to mould round the shape of the tubes, and it is most suitable for the quick collection and transfer of solar heat to the water in the serpentine pipes and so to the solar storage tank. For the purposes of this design, and the spacing of the bent water pipes, 150mm (6in) copper strips will prove more suitable than the next size up, which is double the width. Copper to copper is the best and easiest of thermal bonds but even so the adequacy and accuracy of the bond is of paramount importance. The bond is made by shaping the copper sheet tightly round one half of the pipe and soldering the two completely together.

The copper sheet should be of the 'half-hard' type. The alternatives are—not surprisingly—'soft' and 'hard'. 'Hard' is too difficult to work easily round the tube shape, and 'soft' is so much the other extreme that the parts of the strip that should remain flat will tend to have a dented and rather battered appearance. A thin sheet will be easier to work and easier on the pocket. For the panel described you will need to cut four equal strips from a length of 5.4m (17ft 10in) assuming the suggested width of 150mm (6in). These four strips should be laid over the serpentine tubing and provided with semi-circular grooves using simple wooden blocks as formers (Fig 29).

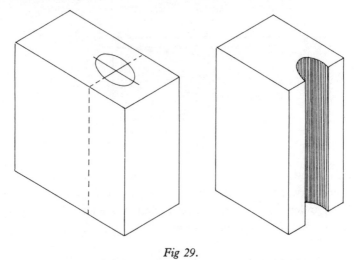

Fig 29.

To make a pair of blocks, take a piece of wood 100 × 100 × 75mm (4 × 4 × 3in) and draw or scribe a line 25mm (1in) above the 100 × 75mm (4 × 3in) face at one end. With the block standing on its narrow side and using a 16mm ($\frac{5}{8}$in) bit, centred on the line, drill a hole right through. Then saw through the block on the line. For simplicity in forming the copper plate round the tubes, cut three such blocks in all to use as three pairs of clamps. First form the plate at each end, leaving the blocks in place. The third pair of clamps should be used in the centre and then worked outward progressively to the end blocks.

When all the forming is complete, the tubes and sheet must be bonded together by soldering. First make quite certain that the surfaces to be bonded have been thoroughly cleaned with wire wool. Then take care to fill in any gaps between the surfaces. The soldering may be accomplished in the orthodox way using flux and solder with a blowtorch. Alternatively, there is a solder 'paint' which combines flux and solder and can be painted straight on to the cleaned surfaces. The blowtorch completes the bonding. Finally the upperside (pipework on sunny side) should be cleaned again and have self-etching primer applied before the matt black heat-absorbent finishing coat.

An absorber made to these directions will have an absorber area of 0.8sq m (8½sq ft). With this knowledge, and an idea of your hot water requirements, you will be able to work out the number of panels needed. A suitable family average to work from is 160 litres per day (35gal per day). It is not particularly difficult to assess the differences in family needs from the average. The main variant is bathing, which may be affected by the nature of work, or the difference, say, between daily as opposed to weekly bathing patterns. If possible, design for a much larger usage than reality may demand so that in indifferent weather there is a good supply of partly warm water. This is cheaper than cold to bring to the required temperature.

A simple rule of thumb for calculating need is to allow 1sq m of absorber area for each 50 litres of hot water wanted (1sq ft per gal). To allow a margin on the safe side would mean two of the DIY panels described for each person. While cost goes up with the number of panels, the extra is minimised when DIY panels are used and the overall economy is the more worthwhile. The next factor to examine in more detail is therefore costing related to solar heating returns as opposed to the ever increasing costs of the orthodox alternatives.

Calculating the cost of solar energy

The user of solar energy stands on its head the old saying that 'where there's muck there's brass (or money)'. He is helping to save diminishing remains of fossil fuel; he is helping ecology, as the air remains cleaner than it would do otherwise; he helps to save fuel imports; his example is likely to be followed by a growing number of like-minded people; and he is looking to easing political problems of the future, for nations are already showing signs of being at each others' throats to corner as much as possible of remaining available mineral fuels. The scale of such benefits cannot be measured: we simply know that they exist.

There is nothing new about energy from the sun, but the cost of collecting it has always been a major snag; it has been reserved for special projects. But now the cost of domestic solar heating (and that for industry and, for example, swimming pools) has gone down. This is partly because of the boost given to the now dozens of manufacturers in this field as a result of the 'energy crisis' which started in the early 1970s. Action by the oil producing countries was the crux; the western, or industrialised, countries found it economic to pursue every alternative avenue for energy sources. While on national scales nuclear energy may have come out on top, to the householder the resultant high price for electricity and the parity demanded, for example, by the gas industry as against electricity charges has made investment in some form of solar 'back-up' much more attractive than in the past. At least to the initiated building their own house, intensive insulation from below ground upwards and the maximum use of, and storage of, solar energy have become musts.

Savings provided by solar energy must, of course, vary considerably according to climate. Reduced to the simplest terms, in a very hot country cool baths and showers are the order of the day, and a quite small area of solar collection panel will make up any difference. At the other extreme, in very cold climates with little sun radiation penetrating, solar panels cannot justify their cost. The temperate climate of the United Kingdom lies well within these extremes, even though—from north to south particularly—there are considerable differences.

The costs of orthodox fuels like gas and electricity—and of course coal in one form or another—go up so quickly and unpredictably that comparisons with alternative energy sources like solar heating are misleading. At any time one may say that a capital investment in an alternative energy source will pay for itself in 'x' years by reducing bills for orthodox fuels. Past bills are a poor guide, however, when orthodox fuel goes up in price so quickly, and interest rates vary. The alternative may well, however, pay for itself in a much shorter period than expected.

The benefit from solar energy in one's own part of the globe remains virtually a constant, maintenance being low over the years. In the south of England, for example, it can be shown that each square metre tilted towards the sun receives something like the equivalent of 1,200kWh—enough to supply 1,200 single bar electric fires for one hour! But one must work to an overall efficiency of about 35 per cent at which each square metre of collector would provide something in the order of 400kWh, or 4/10ths of a single bar electric fire. However, increasing the collector area helps only within reason: on a dull day there is little gain, whereas on a hot day a small collector area will provide more than enough hot water. In practice 4sq m (40sq ft) of collector would provide something like 1,400kWh of worthwhile heat, and 6sq m (60sq ft) would be likely to reach 2,000kWh. If one assumes a 'credit' of 1,400 to 2,000kWh from solar source, then comparisons may be made at the going rate at the time for the cost of the 'back-up' orthodox fuel in use.

In calculating economic feasibility it is easy to fall into the trap which catches so many motorists with their cars—a form of wishful thinking. As with cars so with solar heating, depreciation may so easily be forgotten or given an artificially low figure unrelated to what the real replacement cost will be in, say, three years' time. In addition to initial cost there must be some figure for maintenance: for example, it is more time-consuming and costly to redecorate and protect the panels on the roof than the more accessible gables and guttering. Then there is the interest that might otherwise be gained were the capital cost to be invested. If a loan, or increased mortgage, is needed to achieve the alternative, then the interest must be taken account of in determining the point at which the solar system has paid for itself.

On the other hand money invested in solar heating shows a return in lower fuel bills, and there is no tax on the gain. Again, if an increased mortgage is used, advantage may be taken of further tax concessions. Also, money invested now is likely to be of more value in the long term than money invested after a period of further cost increases and inflation. Finally, there is the factor of adding to the value of your house when the time comes for resale. Increased fuel costs are a major factor in inflation, so that the benefits of curbing them are likely to be of increasing value in real terms.

Recommended reading: Kevin McCartney, 'Practical Solar Heating', Prism Press, Dorset, 1978.

3
Burning Wood and Other Solid Fuels

Although wood may be considered as the original source of additional heat for man, from rubbing two sticks together onwards, the finely controlled, economic exploitation of this energy source is comparatively recent. During the current energy shortage expertise in the use of wood, sometimes mixed with other fuels, has been taken near its limit, though perhaps not even yet to the ultimate in efficiency. Extensive research needs to be undertaken in the supply, nature and costing of wood as fuel.

How do you buy wood for the fire (as opposed to firewood in the sense of sticks for firelighting)? What is a 'load' of wood and how can it be priced? By the size? By the suitability? By the extent of seasoning beforehand? By the weight? If by the weight, what wood supplier is equipped with proper facilities for weighing? And even if weighed, is the wood going to be green or dry, hardwood or softwood? What output of heat can be expected from the load of dry, green, hard or soft? Will it spit? Will it cause resin problems in the chimney? And how, 'at the end of the day', do you calculate value in terms of price and convenience against alternative fuels; or evaluate the benefit from time to time of the pleasant appearance and usually pleasant smell of a wood fire? The householder may make up his mind on the basis of answers to some of these questions.

If any solid fuel heater is to be used, its compatability with wood can in many locations be particularly important. The decision to include wood as a regular source of heat will be to some extent influenced by the location of the dwelling relative to wood supply. The decision may also be affected (in an urban district, for example) by the availability now of solid fuel heaters which will take wood with or instead of alternative fuel.

The use of wood as a fuel must be considered responsibly. Fortunately legislation plays its part in most countries, with the notable exception of the ecologically essential rain forests of South America. England had its real energy crisis in the seventeenth century when whole forests had been ravaged. The situation was saved only by the timely arrival on the market of coal. Total reliance on wood for energy is certainly not practical on a wide scale, for several acres of woodland would need to be properly husbanded to provide all the needs of the average family. But wood, when readily available, can be backed up by—say—solar energy and insulation to make an important energy contribution. Smoke pollution is less than that of other common fuels, and it contains fewer irritants. For example, woodsmoke has virtually no sulphur dioxide. Also there is the psychological factor that to most people the smell of woodsmoke is pleasant, even nostalgic. There is a minimum of ash (which itself can be recycled into fertilizer or soap). According to the US Forest Service, some thirty per cent of a city's refuse is basically wood. Cared for, wood supplies are replaceable, and are available without ecological damage such as mining by shafts or opencast.

In considering wood as a fuel, let us first take the vantage point of the family with some wooded land. A managed woodlot can be assessed in both the short and long term. Long-term considerations include the growing time and suitability for heat output of the type of wood which will grow on the particular land. In the short term there are the trees already standing; among these remain a great many elm which, robbed only of their life-giving bark by disease, are dead but still standing: fine for heating purposes.

The ideal siting of a woodlot would be the windward side of the dwelling because an effective windbreak can reduce heating needs by as much as a third, according to US Government research. A woodlot has some side benefits. For example, animals sheltered from the wind will put on more weight during the winter. Also both food and shelter is provided for wildlife.

Before calculating the long-term heating output from a woodlot one must have some kind of measure. The standard is the 'cord', named after a piece of cord 8ft long. If you were to sink two posts 4ft apart and 4ft high, then a similar pair 8ft away, you would have the volume of a cord. With 8ft logs filling up the space, there would be about 80cu ft of actual wood, the remainder being airspace. In practice few local suppliers of logs will even have heard of the measure, and often will not know the full nature of the wood; that is, how long it has been seasoned and how much—if any—is hardwood, with an appreciably greater heat value in relation to its volume. Nevertheless, the cord remains a useful measure for comparisons.

To give just one example, pruning and picking up broken branches in a woodlot will yield about one cord per acre per year. If you have to grow a woodlot from scratch, buying wood and perhaps other fuels in the meanwhile, then you have an obvious problem. However, trees will often grow on land unsuitable for other crops because of steep slope or perhaps the stumps of hardwood trees felled previously. Quick growing trees will provide fuel in five to six years, but hardwoods take much longer. If earlier clearance of hardwoods has caused erosion on the site, the quick-growing trees such as hardy conifers will help to reclaim the erosion and in time enable hardwoods to be grown successfully once again. Of course, trees do not need to be full grown before being used as fuel, and they have to be thinned out from time to time, thus providing a moderate supply.

Sources

The Forestry Commission has no guaranteed supplies, but if you are within reach of one or more of their very many sites, it is worth asking. Demolition sites should be fruitful for the price of a drink or two as most waste timbers are burned on site simply to get rid of them. Take care, however, not to transfer dry rot to the timbers of your own house. If you live in the country, it may pay in the long term to buy a chainsaw, as there is a variety of wood sources. They range from the ravages of tempest to people who want a tree removed and will let you have the wood for felling it and taking it away. A word of warning here, though. Thousands of people use motor-driven chainsaws safely but they are potentially dangerous. Follow makers' directions *implicitly*. *Never* attempt to fell a tree until you are well used to using the saw, and never let the chain get blunt so that you have to use effort in cutting and lose your balance. Make a point of finding a regular user to stand by while you practise, and never be out of earshot of someone else when working, in case of accidents.

Heat value

Broadly, the heat value of wood increases with the hardness of the wood itself. In other words, oak has more heat value than pine; in the USA hickory would top the list. Again, dry wood will give more heat, weight for weight, than green wood. If you start with green wood and dry it, you may find value for money balances out because of the reduction of the wood's weight during drying. When wood is drying, it needs some form of shelter; but it should be open for maximum air circulation, and the logs piled to allow easy airflow. In the UK elm and ash are good for wood fires; both have a high heat content. If combustion is uncontrolled, however, ash may burn too quickly; while elm needs a good fire going before it will ignite properly. Elm is at its best in controlled combustion stoves or (not unfailingly) for staying in all night even on an open fire.

There are other factors to consider apart from burning speed and ease of ignition. If you split your own logs, a wood with a straight grain will save time and effort. Conifers and ash are easy, elm and sycamore more difficult. Sparks can fly from some woods, including most of the 'soft' variety—particularly when green. The pressure of pockets of water vapour builds up to bursting point with the heat. This tendency is less when the wood has been air dried and is of little consequence in a controlled combustion stove with the fire not directly open to the room. Even when certain woods, such as larch and spruce, are dry, their resin content may still offer danger on an open fire. The combination of moisture and resin can also cause extra smoke and leave ignitable soot and creosote in the chimney. Very slow, controlled combustion of such woods should alternate with a really hot fire from time to time to reduce the risk of build-up leading to a chimney fire.

With open fires, or those which can be exposed to the room when required, wood from fruit trees is especially pleasant. Apple is the favourite for its flame and beautiful fragrance. Well dried driftwood also has a good flame. A cheaper kind of 'wood' now popular is a log pile made from waste paper. Old newspapers and magazines are soaked, then wound tightly round a slim cylinder in one of the 'log making' devices, fastened quickly with the ties provided, and the cylinder withdrawn to allow a central air passage. Alternatively, the 'logs' can be tightly rolled by hand. In this re-cycling the tight paper is in effect being returned to its original nature—wood.

Heaters

As the best dry hardwood will have at most only half the heat value of the same weight of coal, few people can be self-sufficient for all heating purposes. For a typical family the woodlot would have to cover several acres, as mentioned, as each acre would produce about one cord per year over a period. Therefore to heat the house and water it makes sense to get a stove that will take a mixture of wood and fossil fuel, ideally any type of solid fuel separately or in any mix. Typically it might be used as a cheery wood fire on long, cold winter evenings, then given coal or anthracite at bedtime.

Some of the older type of heaters are illustrated. One example is the Doughnut (Fig 30) flue, which had obvious advantages in its day. However, in the best modern burners there is no need for the stovepipe to be a source of heat, as the hot gases are channelled round within the metal cladding of the heater before reaching the flue, giving out their heat in the process. Old soldiers (and many not so old!) will recall the oldfashioned barrack room or Nissen hut stove literally glowing red hot in some areas. But safety codes with children and the aged in mind now require that the surfaces should not be hot enough to cause a burn. They tend to act more as large, powerful convectors than as radiators.

In some stoves one can have the best of both worlds, using an opening front for radiant heat and sight of the fire, whenever one likes. But remember that the efficiency of a controlled combustion stove will be reduced while the door is open. A pioneer was

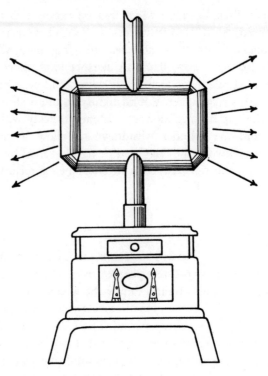

Fig 30. The 'Doughnut'.

Benjamin Franklin, whose stove, designed in the early nineteenth century, had folding doors at the front. The Franklin (Fig 31) went a long way towards controlled combustion, and is still selling at a comparatively low price in the USA, particularly for the 'log cabin' type of holiday home.

Charcoal is used in some countries, notably Japan, with different fireboxes for different purposes. One burner can be used for warming the hands, another type for the feet and yet another for bed warming. But apart from fire danger there is real risk of carbon monoxide (CO) poisoning. Therefore people have to remain warmly dressed for winter to withstand the essential ventilation.

Normally heaters that will take wood have manual controls for draught. While they can be considered as virtually complete combustion burners, there is still waste because the occupants of the room begin to feel too warm before being prompted to adjust

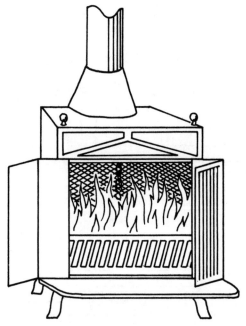

Fig 31. The famous Franklin—still going strong.

the controls. Entering the market at the time of writing is an interesting newcomer called the Pembroke (Combined Energy Consultants Ltd, Llanelli, South Wales). This incorporates a sensor of room temperature and a thermostat which keeps the burning rate steady according to the preset heat output required at any time. As a space heater for any size of domestic room, it is an economical convector with a fire visible through a magnifying reeded front that simply lifts out of sight when an open fire is preferred. It can be fitted with a back boiler for hot water and has provision if required to heat up to six radiators. Loading is from the side and can be of any wood or mixture with fossil fuel. Stoking is normally required only twice in 24 hours using mixed fuel. The thermostat avoids the hit-or-miss problems of manual control.

Chimneys

The standard type of fireplace, chimney breast and chimney costs a lot of money, which is one reason why so many houses or flats

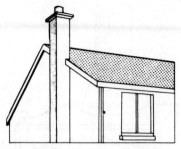

Fig 32. An inexpensive concrete block chimney added to a previously chimneyless house.

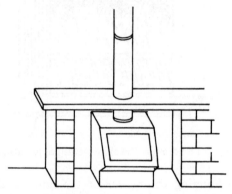

Fig 33. A stovepipe of modern design to provide a chimney. It can be disguised by almost any type of cladding, and upstairs simply boxed in (Fig 34).

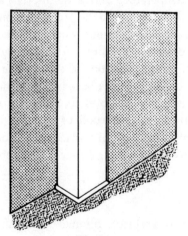

Fig 34.

are without them. However, there are two ways of putting up a ready-made chimney at a much lower cost. The first is of concrete blocks with an appropriate hole in the centre (Fig 32). The other uses a modern stovepipe (Fig 33). Both can have a simulated chimney breast for appearances, and be boxed in simply on the first floor (Fig 34). Whatever heating unit you use, the manufacturers will be able to advise on sizes, but for solid fuel fireplaces a guide is given in Fig 35. Cladding or rendering can be to your own choice.

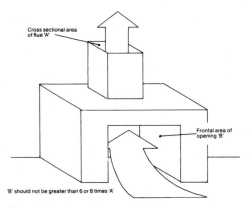

Fig 35. Basic requirements of hearth design.

4
Windpower

The use of winds and tides goes back almost into prehistory, but before the advent of steam power windmill designs had become as advanced as their variety might suggest. After a long period of neglect, the energy crisis has caused a rethink about the use of wind. (The same applies to tides but they are not in our brief.) Wind energy worldwide, at ground level alone, would be about double the energy of all the power stations ever built. Of developed countries, Denmark, Great Britain and—of course—Holland are among the most windy. Their supply of this energy source is fairly constant, but with the advantage that windpower is at its peak when solar energy is at its lowest. Broadly, three-quarters of the wind potential energy comes between November and April. The question is how to capture enough to supplement our individual needs or, with the aid of solar energy, satisfy our needs.

While in the past wind has been used to power mills or other direct-drive machinery, nowadays the concern is to turn windpower into electricity. Some of this electrical power can be used direct but storage in one form or another is also needed. The storage is most commonly in batteries but it may also be transferred to warming or prewarming the hot water supply for more general benefit.

Other things being equal, the electrical output in a given wind condition will depend upon the diameter of the blades, and in this context it is important to understand that wind speed increases this output by the cube of the wind speed itself. That is, if the wind speed doubles, the potential output will be eight times greater. This means close control of the speed of the blades and the nature of the generator—probably an alternator—to cope with greatly varying wind speeds. So the next step is to evaluate three things. First, the *average* wind speed; secondly, the size of the windmill; and third, the capacity that the generator will stand without fear of burning out.

An important aspect is safety. A large windmill in a gale presents a great hazard—indeed a frightening one—unless there is proper control of the braking mechanism used and the mounting is secure. If buying a commercially produced windmill, it is best to ensure that the manufacturers guarantee to supply a suitable tower and that this, with the help of its guy-rope type supports, will survive any wind force possible in your terrain.

The particular conditions at your site are technically described as the 'wind regime', and it is especially important to know what it is between the end of September and March. It is possible but expensive to buy or hire an anemometer (wind speed indicator) and take regular readings to get an average. However, it is easier, if not so accurate, to get the opinion of your nearest meteorological office. In the west of the UK wind speeds average about 24kph but in the Midlands 16kph.

The site for the mill is the next consideration. At the top of a hill—no problem. At the bottom of a hill conditions are still quite good on the windward side of prevailing wind direction, if there are no obstructions likely to cause turbulence. Half way up on the leeward side is the worst site, as the wind will probably be deflected over the top of the blades. Turbulence should be avoided if possible because of the disproportionate stress it puts on the structure. Not only is there a varying wind direction causing the blades to swing to face it but also varying pressures. Even a small generator suffers much more strain than a static structure like a television mast.

Choosing a windmill

There are two basic types of windmill. The most common through history is the mill with vertical sails or blades and a horizontal drive shaft. The other has a vertical shaft with the blades spinning horizontally, like the 'mobiles' often seen in shops with the advertising material swinging round in any draught. Of the horizontal drive mills the most familiar are of the traditional type with slats or sails, mainly associated with the Dutch landscape. These have proved efficient at pumping water or grinding cereals

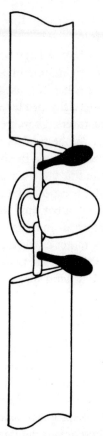

Fig 36. Bobweights used to 'feather' the blades of a mill when windpower reaches too high a speed.

into flour but are not favoured for driving generators—or at least without means to gear up the speed of the drive shaft. In terms of, say, the car engine they produce high torque at low speeds (analogous to the diesel engine) whereas the generator should be spun quickly even when wind speed is low. Gearing up the slat mill adds to the cost and increases losses by friction. Blades on horizontal drive mills now mainly look like aircraft aerofoils. Some years ago the Danish State Electricity Board built a three-bladed version which feeds 200kW into the national grid.

An important advantage of the aerofoil type of blade is that it can be 'feathered' automatically if wind speeds become dangerously high. That is, the angle of the blades can be turned to alter the pitch, as with aircraft propellers. The most popular way of making this important safety precaution automatic is to use bobweights (Fig 36) which with increased rpm swing out and reduce the blades' angle to the wind. With a correctly designed blade, the speed at its tip should be about six times air speed, so a blade of 1.8m dia in a wind of 40kph should give a generator speed of 700rpm without gearing, which is just about right for most generators. Ideally the bobweights will be adjusted so that, wind permitting, the selected generator will be running at its optimum.

The history of really large mills intended to add their output to any national grid has not been happy. But mills are commercially available with an output of up to 6kW, which is enough to meet the needs of a family or farm. The problem here is the high cost, although apart from getting it back from lower fuel bills over a long period there is the attraction of being independent of power cuts, which always seem to happen—particularly in the countryside—in seasons when a cold wind is in plentiful supply.

Vertical shaft drive

A basic advantage of having a vertical drive is that it avoids the problem of 'luffing'. These days this old seafaring term is applied to machines like cranes which have to turn on their horizontal base. If the blades drive a horizontal shaft, then almost certainly the generator will be mounted on that shaft, and with the propellers always turning to face the wind, there is the problem of getting the current down to the ground. An ordinary cable could get horribly tangled, so 'slip rings' are used. These have sliding contact to connect with the vertical shaft and base. Problems of dirty contact, weather conditions, and so on, mean that slip rings can be sources of trouble. With a vertical shaft drive there is none of this cost and potential trouble.

To have a vertical shaft drive to the generator means that the surfaces facing the wind must spin like a Catherine wheel firework mounted in the *horizontal* plane. The generator is mounted at the bottom of the vertical drive shaft and the need for slip rings or any other alternative is obviated. The most obvious snag to this system is that if two blades, one on either side of their mounting, are both facing the wind at the same time then there is an impasse. Again, if one blade starts turning, then the other will be coming into the wind and therefore reducing the efficiency drastically.

However, there are a number of solutions to the problem. One is to make the blades like Venetian blinds, but closing against wind pressure and opening to trail in the wind when swinging to face it on the opposite side of the axis.

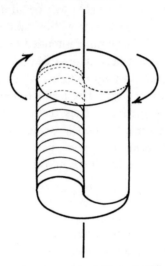

Fig 37. A Savonius mill (vertical shaft drive).

Another method is to have 'sails' which gather the wind when facing it but are so curved that little resistance occurs when they have their back to it. The most famous—and most simple—of this type was the brainchild of Sigurd Savonius of Finland who, ignoring sails, built the metallic Savonius rotor (Fig 37). This design reminds one of a famous racing driver of the 1930s who was his own mechanic. The secret of his success, he said, was to 'add lightness and simplicate'. There is nothing more simple than the Savonius mill. It is usually made of two oil drums slit down one side and then bent so that one slot faces the wind while the other has its back to it. The beauty of this design is the added advantage of an aerodynamic effect called the Magnus principle. This means, in effect, that as a convex face is to the wind the air-flow moving over it has some forward suction, so that it is drawn towards the wind, increasing the effect of the wind on the 'open' drum. Both drums are therefore in unison in getting maximum effect from the mill—transmitted down the *vertical* drive shaft to the generator.

An advantage of the vertical shaft mill is that wind direction is of no account. Unlike horizontal shaft mills it has no need to turn to the wind, as one aspect is always to the wind from any direction.

The other best known type of vertical shaft windmill is the Darrieus, named after its French inventor. It is also known as Catenary because of its profile when operating. (If one suspends a slack chain from two points of equal height the curve it assumes because of gravity is catenary.) This mill has slim aerofoil section blades (usually three) which, owing to centrifugal force when spinning, take up the overall profile of a rugby ball on end. The only strain on the blades is tension, so a stiff construction is unnecessary. Indeed the whole design makes for low-cost efficiency, giving a high output for a given mass of structure. One made in Canada is displayed at the Centre for Alternative Technology near Machynlleth, North Wales, which has a rated output of 3kW in a 23mph wind at 110V DC.

The one snag in this design is that it is not self-starting. The 15ft DAF model at the Centre incorporates a $\frac{1}{2}$hp starter motor. However, a simple and cheap solution is incorporation of a small Savonius rotor at the base to start it going. Speed rises until the catenary blades are swung out; then speed quickly increases.

The future for windpower

At the time of writing the most interesting mill under investigation in the UK is at Reading University under a team lead by Dr Peter Musgrove. This is a vertical shaft type with an 'H' shaped piece at the top. The centre of the bar in the 'H' is mounted on the vertical shaft, while the two verticals are very thin aerofoils of untapered section and perfectly straight. This is simpler than other designs to date and it also incorporates an effective, simple feathering device.

Most of the research in progress for big mills capable of feeding into the national electricity grid is going on in North America

(USA and Canada) and Denmark. But problems are great: what happens in a tornado? So it seems that the most promising developments will be in windmills to provide for a single dwelling, a farm with outbuildings, heat for commercial greenhouses, or possibly a group of houses. So far, for obvious safety reasons, the use of wind is practicable only in country areas, with the exception of small units feeding batteries for emergency lighting during mains power failures and other limited applications.

If this holds true, the research going on in the UK by private firms and individuals is certainly practical. Apart from the work done at Reading University, there is an interesting and unorthodox windmill, suitable for DIY application, developed by Robert Walker in West Wales. This, instead of having the vertical blades kept at the front (facing the wind) has a cone pointing into the wind with the blades at the open end of the cone. As the wind strikes the cone, it increases velocity (an analogy is the Venturi) and impinges on the blades to greater effect. The standard model has a 500W output, adequate for all lighting purposes, with the daily surplus going to back-up water heating. (Combined Energy Consultants Ltd, Llanelli, Dyfed.)

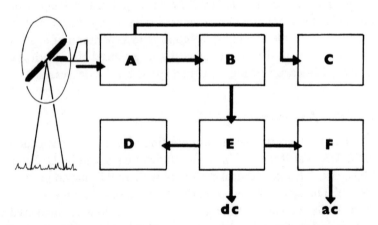

Fig 38. Sample windpower layout: (A) Voltage regulator; (B) Battery bank; (C) Heat storage; (D) Stand-by; (E) Control; (F) Inverter.

An important factor with wind power is the storage of energy for use in periods of calm. Also there is the question of voltage.

Normal electrical storage will be in orthodox lead-acid batteries. They are expensive, but the most economical—because mass-produced—are heavy duty 24V commercial vehicle batteries. Capacity to provide, say, 1kW continuous average for two days would amount to 48kWh, so using commercial vehicle 24V units of 5kWh, one would require ten to be sure of evening out the electrical supply. A wind powered layout is shown in Fig 38.

It is also possible to get small immersion heaters of low voltage to feed surplus energy into preheating domestic hot water. For the future of this application—like that for electrically driven vehicles—one hopes that the search for a low-cost, large-capacity battery will in time be successful.

5
Waterpower

Even though the use of water as an alternative source of energy is confined to those whose dwellings have a stream or are beside a river, the appeal is widespread: perhaps one's next house will be in such surroundings! While there is usually some minor snag with various energy devices, such as cost, safety, and so on, the snag with waterpower demands attention at the outset. This is that the local water board, river authority and anyone else with a say-so must be consulted first. Taking power from the water at one point may adversely affect other established users lower downstream, for example. The water may pass through your property, but that does not necessarily mean that you can do what you like with it.

The instinctive appeal of waterpower overrides such problems, as well as the complexity of efficient usage—far greater complexity than would be popularly supposed. Waterpower is another form of solar energy, as it is the sun which starts the cycle by evaporating water from oceans and lakes, and warming the air to lift the moisture. The 'head' of water needed is therefore provided primarily by solar action. A practical appeal of waterpower is its efficiency in a well designed set-up. From 80 to 90 per cent efficiency may be obtained, compared with 25 to 45 per cent from other alternative power sources.

There are two basic types: the familiar water wheel and the turbine. The latter drives a generator, while the former may be used with mechanical connection—to grind flour, for example—or geared to drive a generator like the turbine. Before opting for either of these types, or deciding between several basic forms of water wheel, the first step is to calculate the power available from the water supply. This is calculated as the rate of flow of water (measured in volume per minute) multiplied by the 'head' or vertical drop (in linear measure). These two measurements will tell you whether or not the power is sufficient as a usable source of energy.

Waterpower systems usually require a dam, even of a simple kind, to provide constancy in the power supply. It provides a means of regulating the water flow and, by making it deeper, can add to the height of the water and so provide more head to operate the wheel or turbine. Water wheels and turbines both give their power in a shaft (as torque, or 'twisting power'). This power can be used for purposes already mentioned, but also to drive compressors or pumps.

Water wheels can now be bought inexpensively in plastic for self-assembly but they are still comparatively slow-turning and therefore call for the high cost of gearing up to shaft speeds suitable for driving an electric generator. Water turbines are best suited for driving electricity generators. They are of small diameter but give fast rpm when driven by the available waterpower through a pipe or nozzle. Quite small turbines can provide a worthwhile electrical output. Apart from the turbine or water wheel, the main cost will probably be in gearing, possibly by belts and pulleys if cheaper and appropriate, and the electrical generator itself. Then comes the problem of the dam. While many components of the complete 'power station' can be bought secondhand, the dam is most likely to be of your own construction. With any luck you can use local materials, but the cost obviously goes up if you have to use reinforced concrete. Another variable is the pipe or waterway to take the water to the wheel or turbine. Bear in mind that water is heavy, so that if any kind of viaduct has to be built, it must be strong.

Measuring the waterpower

A complete study of waterpower available, relating it to ultimate use, is too skilled and specialised to be within the scope of the normal domestic needs of a dwelling with the stream or river adjacent. However, some measure of the power available is a prerequisite. You must know the flow and the head.

Calculating the volume

Measuring the volume of flow is the most difficult task, but not insuperable by the practical man. The volume is found by measuring the capacity of the stream's bed and the flow rate of the water. Both normal and minimum flows must be measured. The flow naturally varies not only from season to season but even from day to day, so it will pay to take many measurements over as long a period as possible.

An easy way to measure the flow rate of a stream is to build a temporary dam and to channel the water so that it can be captured in a container of known volume. Time how long it takes for the container to fill and then use the following equation to find the flow rate:

$$Q = \frac{V}{S}$$ where Q = flow rate in cubic units per second
V = volume of container in the same units
S = filling time in seconds

If the stream has too great a flow for this method, you need to make a weir instead of a dam. The weir need not extend right across the stream or small river but can be a rectangular space for water flow in the middle of a small dam. The space must be of known dimensions. The next stage is to measure the depth of water going over the weir—or gap in the dam—the width already being known. The standard table makes the volumes of flow easy to calculate. The depth of water over the centre section of the weir is shown as 'H' and the flow value as 'Q', in this case per foot of the opening. Keeping to measurements in feet, the table is used in the following equation for flow rate:

$$Q = T \times W$$ where Q = flow rate in cubic feet per second
T = flow value in cubic feet per second *per foot of weir width*, from the standard table
W = width of weir gap in feet

For this calculation to be accurate some precautions must be taken. Before starting to build a dam, measure the depth of the

Flow Rate Table

Table for rating the flow over a rectangular weir. Flow (Q) is in cubic feet per second per foot of width of the weir. Multiply the flow value given times the width of the weir in feet to find the actual flow rate.

H, head (feet)	Q, flow (cfs)	H, head (feet)	Q, flow (cfs)	H, head (feet)	Q, flow (cfs)	H, head (feet)	Q, flow (cfs)
.05	.037	1.05	3.51	2.05	9.37	3.05	16.66
.10	.105	1.10	3.76	2.10	9.71	3.10	17.05
.15	.193	1.15	4.01	2.15	10.05	3.15	17.45
.20	.297	1.20	4.27	2.20	10.39	3.20	17.84
.25	.414	1.25	4.54	2.25	10.73	3.25	18.24
.30	.544	1.30	4.81	2.30	11.08	3.30	18.65
.35	.685	1.35	5.08	2.35	11.43	3.35	19.05
.40	.836	1.40	5.36	2.40	11.79	3.40	19.46
.45	.996	1.45	5.65	2.45	12.14	3.45	19.87
.50	1.17	1.50	5.93	2.50	12.51	3.50	20.28
.55	1.34	1.55	6.23	2.55	12.87	3.55	20.69
.60	1.53	1.60	6.52	2.60	13.23	3.60	21.10
.65	1.72	1.65	6.83	2.65	13.60	3.65	21.53
.70	1.92	1.70	7.13	2.70	13.97	3.70	21.95
.75	2.13	1.75	7.44	2.75	14.35	3.75	22.37
.80	2.34	1.80	7.75	2.80	14.73	3.80	22.79
.85	2.57	1.85	8.07	2.85	15.11	3.85	23.22
.90	2.79	1.90	8.39	2.90	15.49	3.90	23.65
.95	3.02	1.95	8.71	2.95	15.88	3.95	24.08
1.00	3.26	2.00	9.04	3.00	16.26	4.00	24.52

stream. The depth of the rectangular opening in the dam, ie the weir, should be the same. The opening in the dam should have its baseline at least a foot above the downstream water level. The weir opening, apart from being centrally situated, should have a width of *at least* three times its depth. This will ensure that it will be able to cope with high flows yet will not be too large for low flow levels to be measured.

The three edges of the weir opening should be at an angle of 45°, with the sharp edges upstream. This avoids turbulence and therefore aids accurate depth measurement. As the dam need only be temporary, it can be made of any suitable material that comes to hand, such as logs sealed with clay, tongue and groove boards, supported iron sheets, and so on. All water must flow through the weir gap, so leaks must be sealed, including those into ground not previously carrying water. Polythene sheet is useful, but use your own judgement about when it should be laid during your dam construction. Big streams and rivers require alternative methods of measurement but these are beyond the scope of the farm or house dweller.

Calculating the 'head'

There are several ways to calculate the head, or pressure, of water. Most of the measuring systems are probably beyond the scope of the potential user, for the head does not mean the actual drop from one level of the stream down on to the water wheel or turbine, but the total drop of water from a distance. In the UK the large-scale Ordnance maps, with their clear contour lines each with stated heights, will be a help. With luck you may also get help from the county surveyor to settle on a reasonably accurate figure for the 'head'.

Having calculated the flow and the head, it becomes possible to work out the power of the water source. The power, whether or not to drive a generator or some mechanical power transmission, is usually worked out in horsepower. The basic calculation is:

$$THP = \frac{Q \times H}{8.8}$$ where
- THP = theoretical hp
- Q = flow rate in the units used —say, cu ft per sec
- H = head in ft
- 8.8 = correction factor (for friction and other losses)

Friction losses are mainly in water pipes feeding the prime mover, so the selection of such pipes is important, even though much depends on the type of wheel or turbine chosen for the job. The main decision will be between wheel or turbine, and the main factors influencing your decision will be:

Flow rates, min and max. The 'head'. Type of soil (resistance to or likely effect of erosion on hitherto dry ground). Pipe length. The water itself (cleanliness, acidity, etc). Tailwater level below wheel or turbine (min and max). Possibility of freezing. Available materials. Cost. Maintenance.

Water wheels

Undershot

Probably the forerunner of all water wheels, it began as a simple paddlewheel lowered into the water flow until the blades turned. Such wheels of 14m diameter powered the fountains at Versailles with an efficiency of 10 per cent. Improvements were made in water control to increase its speed as it hit the blades; at the same time greater quantity reduced interference from the backwater. The main loss of efficiency arose from the turbulence as the water hit the flat paddles. The curved blades of the Poncelet wheel took this basic design to its ultimate development (Fig 39).

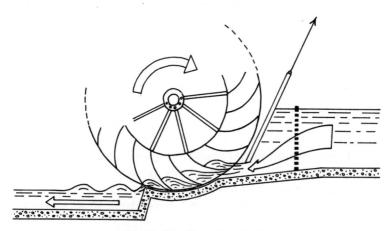

Fig 39. The Poncelet wheel.

The Poncelet wheel normally operates at low rpm but gives high torque. It is suitable for a head of one to three metres and for a flow ranging from enough to turn the wheel up to whatever flow is available. In a DIY application the spokes are normally of heavy timber with the vanes—or 'buckets'—usually formed from sheet steel. The curve of the vanes reduces shock, which reduces energy output, and the water runs up, giving impetus to the wheel before it falls back with only gravitational energy into the tailwater. Properly designed, the Poncelet will give efficiencies of 70 to 85 per cent. The diameter preferred is about three metres minimum to an upper limit of four times the head.

Of importance to the efficiency is the snug fitting, usually of concrete, under the base of the wheel. This is to keep the water in the vanes or buckets, but it must not touch the wheel itself. The course of the tailwater immediately after the wheel must be deep enough and wide enough so that the tailwater can escape quickly to avoid impinging back on the wheel.

Overshot

As an alternative to the undershot wheel, the overshot comes into its own for higher 'heads', or falls of some three to ten metres. The efficiency is similar to that of the Poncelet wheel if the overshot type is equally well designed. With an overshot wheel there is an energy gain as the water hits each bucket, but the main source of power is in the weight of water in the buckets as they descend. The buckets must be shaped so that the water enters without wasteful turbulence so that the buckets are as full as possible for their power-supplying ride down. In a dry summer there will probably be less water to work the wheel, so total energy output is reduced; but to some degree this is offset by the extra efficiency, as spillage is reduced when the volume of water fails to fill the buckets right up. As with the undershot wheel, care must be taken to guide the water just past the centreline of the wheel and fill the buckets without loss of water to the sides.

When considering a water wheel, always have in mind that they are best suited—via belts and pulleys, for example—to drive pumps or light machinery. If used for electricity, there is not only the quite high cost of the gearing up to suit an electrical generator, but also the problem of the changes in speed of a low revving prime mover.

Water turbines

These have the advantage of turning at high rpm and so lend themselves to the production of that versatile energy source—

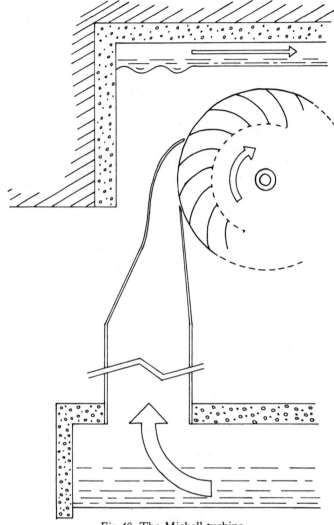

Fig 40. The Michell turbine.

electricity DC (direct current) or AC (alternating current). Nearly all household electricity supplies are AC, but one's own DC supply can be used equally well for one's own outlets and equipment designed for it. There are two basic types of turbine, impulse and reaction.

Impulse turbines have vertical wheels which use the kinetic

energy of the jet of water on their specially shaped 'scoops'. These are intended to turn the water through 180° if possible, and at speed the water then falls to the bottom housing, which contains little more than air. This type is often called the Michell and sometimes the Banki; the English engineer A. G. M. Michell presented the first Paper on the subject, and Donat Banki of Germany later contributed a further significant Paper (Fig 40).

Reaction turbines, in contrast, are horizontal wheels encased in a water-filled housing, and the outlet below causes a little vacuum which increases the total head and reduces turbulence. Impulse turbines of the Michell type can be fabricated by the DIY man but the reaction type has to be 'bought in'. The Michell type is not difficult to construct, although some welding is needed as well as some basic machining. Both the steel and piping required can be obtained from stock sizes, and the design covers the range of flows and heads likely to be harnessed by the user. One advantage of this type is that if there is variable flow the maximum efficiency will remain the same—given a constant head—from a quarter of the design flow up to the maximum.

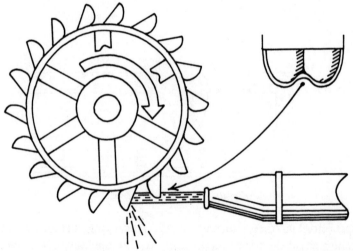

Fig 41. A Pelton wheel, showing (inset) the cut-out in each bucket made to avoid fouling the jet.

The other type of impulse turbine accepted for its efficiency up to and even exceeding 90 per cent is the Pelton wheel (Fig 41) which, if there is enough head, will develop high rpm suitable for generating AC power. A Pelton wheel, normally small in diameter, has shaped bucket cups round the diameter which are turned by a high-speed jet of water from a special nozzle. As a high head of water is required, the device is in its element when driven by water from a mountain stream with a head upwards of 150ft. It will then turn at up to 1,000rpm, needing little gearing up to suit generator drive. In calculating flow and head, as already described, the minimum head approximation should allow for a head loss of one third for friction in the long pipe run. This figure of one third arises from balancing efficiency against capital outlay, for to reduce friction loss further, the outlay would tend to exceed the benefit of the power provided.

Power to electricity

There are real problems in converting waterpower—and windpower—into AC electricity, partly because of the high rpm needed and the loss of efficiency overall in reaching such speeds. Also the AC generator requires consistency of drive. So in small-scale generation DC has more going for it. Alternators used for vehicles, complete with their voltage regulators, will work from about 700rpm upwards, and standard vehicle-type DC batteries will look after storage. Many vehicle systems can be adapted for household use, powering lighting, radios, tape recorders and cassette players. Suitable television sets can also be used, and surplus energy can prewarm water. Again, gearing up can be achieved with gears adapted from scrapyard back axles. One drive shaft must be locked to pass power through the other, but ratios of 3 to 1 and even up to 8 to 1 can be obtained.

The DIY aspects of water wheels and turbines are given here for the comparatively few people who have a waterpower source now. For the calculations in more depth and a list of further reading a good reference is *Energy Primer*, collated by the Portola Institute and available in the UK through Prism Press at £4.95.

6
Methane Fuel

Although methane production is not applicable to the urban household, it does have real interest for communities and for farmers. For example, the Greater London Council has a sewage plant that produces some 164,000 cubic metres of gas *every day*, which adds up to a fuel value of 40 million litres of petrol a year. Where municipal plans are put forward even small households need to be taken into account, for reduced fuel demands can reduce or help to hold steady the money paid in rates levied by local authorities. The disadvantage of small-scale urban production is well stated by T. P. McLaughlin in *A House for the Future:* 'Think of the matter from the point of view of your neighbours—not many people relish living next door to a gasworks, and how do you feel living next to an *amateur* gasworks ?'

Methane is a product of decayed organic material broken down into different easily identifiable substances destined by nature for reuse. The organic waste can be decayed with or without the presence of oxygen (aerobic or anaerobic). In either case the end products are desirable; but anaerobic gives us both methane and excellent fertilizer. Frequently the decomposition of organic material has taken place under water or in what are now subterranean caverns. Marsh ground is a common source, hence the alternative name of 'marsh gas'. In these conditions the gas may bubble to the surface and if ignited by the sun through a glass chipping, or by lightning, may continue to burn; the flame, hopping from one bubble to another, gives the will-o'-the wisp effect. Subterranean sources have provided the supply from the North Sea. The heat value of this methane is higher than 'town gas' produced from coal. There have been serious explosions in some refuse tips when big gas bubbles have formed below the penetration level of oxygen, and sewage plants can produce unwanted methane if not adequately aerated during processing.

Gas production takes place in a digester under control. Sewage, whether it is in the form of bird droppings, animal manure or human waste, speeds up the production of gas, but it must not come from any source containing antibiotics, since these may kill the bacteria that set off the production process. The same applies to detergents, a fact that many countrydwellers discover when using septic tanks for purification of waste matter from the W.C.

As an individual producer, the mixed farmer is in the best position to benefit, for he can turn animal waste into gas and make use also of the quality fertilizer. Even so, the usefulness will depend upon the supply of waste vegetation, including wood chippings, straw, even paper and rags of 'natural' manufacture, but not manmade fibres or plastic. The relative value of manure from farm animals, if a laying hen is taken as 1.0, is approximately 2.5 turkey, 5–20 sheep (lamb first), 23–40 pigs (feeder first to breeder second), 50–70, 120–150 and 180–200 horses (in order pony, medium and large), and finally bovine with the ascending order of beef stocker, feeder, dairy heifer and dairy cow in the proportions of 120–150, 150–200, 250–300 and 300–350.

'Bio-gas' is commonly used as a term for this type of gaseous product. Most bio-gas in this context is made up of 64–72 per cent methane, 24–32 per cent carbon dioxide, the remainder being hydrogen, nitrogen and other gases. A common ingredient is sulphur dioxide, particularly if animal offal is included in the mix to speed production. This 'rotten eggs' smell is usually quite light and it has the advantage that in any sophisticated system the detection of leaks of potentially dangerous gas is made easier.

The crucial question is, does a digester pay? This depends mainly on the availability of sufficient quantities of vegetable matter and the manure to help provide the essential bacteria to get the process rolling; also whether the cost can be covered both of construction and of maintenance of input and output and the ancillaries? On a large scale, the answer is clearly yes, and community schemes should certainly be supported. The urban house is virtually ruled out, for obvious reasons. What of the farmer with a promising mixture of hens/animals and crops?

There is a third benefit for the farmer, in addition to methane and fertilizer. This is disposal of waste. With this can be linked the claim that the fertilizer is superior to the use of dung in the ordinary way. But it would take the daily droppings of 60 chickens to produce enough methane gas to cook plain meals for one person for one day. Conservationists and those who wish to live off the land almost exclusively for their essential requirements—as is the ambition of the Alternative Technology Centre at Machynlleth, in Wales—will find a methane digester most valuable as part of an integrated energy system. The biological needs of digesters for heat can be met by solar or wind sources, while their waste products can be used as fuel. (The warmth required for speeding the cycle of a methane digester can be provided by the digester itself if large enough, but otherwise needs a little energy help in heat for the best overall efficiency.)

One fairy tale can be dismissed. You cannot run a car by having a chicken in the luggage compartment. Methane has a higher octane rating than is needed for your car and is therefore probably best used for a compression diesel engine near the digester: again one comes back to integrated systems. Waste heat from the engine can be fed back to the digester. Another basic problem with methane powering a mobile engine is the space it takes up. It can be compressed up to a point in cylinders located in the car, but the compression equipment costs money and the mileage range is still not great. In properly integrated systems on large mixed farms, however, and especially in community schemes, methane is attractive.

7
Heat Pumps

The heat pump is conventionally described as a refrigerator in reverse. The latter is usually stocked with food at room temperature, from which heat is extracted and returned to the kitchen. The basic principle is straightforward. A fluid (which can be a liquid or a gas) is compressed, either by being expanded by heat or by being compressed by a power-driven compressor. Just as a bicycle pump gets hot in use, so the compressed fluid gets hot and this heat is then dissipated in a radiating coil. As it expands, it cools—in this case in the refrigerator, thus cooling the food. Heat is extracted from produce at room temperature down to below freezing point if required. This cycle is continued under the control of a thermostat.

The heat pump now becomes easier to understand. It turns the system round by removing the heat from a considerable amount of what is called 'low-grade' heat and reissuing it as a smaller quantity of usable heat. There are many obvious sources of low-grade heat, ranging from discarded bathwater to the ground in your garden. The latter, even in a cold winter, will be above freezing point at an appropriate distance down from the surface. Heat from the sun is another source, although there are storage problems. As for bathwater, the total heat wasted may exceed that of a domestic boiler at its normal thermostatic setting.

Fuel wastage is extraordinary. When measured accurately, the efficiency of most machines is really low. A steam engine may give only 15 per cent efficiency, much of the waste going into cooling and supplying surrounding air with heat. Heat loss is applicable to most engines, including electric motors, which supply only some five-eighths of the input, and the input has already been 'devalued' in turning the original energy source into electricity. This is particularly true of electrical waste, for although the power stations are collectively the biggest consumers of raw energy, only one-third reappears as usable heat in homes and factories. In the UK some 60 million tonnes pa of coal is lost in cooling towers or cooling water, the latter warming rivers to no purpose. Some power stations in other countries are more conscious of this absurd waste and feed the heat to housing estates, and so on. Even so a great deal of original energy goes to waste.

This is where the heat pump steps in. It takes this waste—or 'low-grade'—heat and turns it into a smaller quantity of usable heat, useful heat. The heat pump is specially useful in an industrialised country, as there is so much low-grade heat available from a variety of industrial processes. In the heat pump the low-grade heat is cooled in the expansion stage by passing the air or water over coils, or by putting the coils in the ground beside the building. Extracted heat is passed to the compression stage inside the building. In a house the compression coils can have a fan to circulate warm air via ducts. In this type of application the heat pump is more efficient than its counterpart, the refrigerator. In the latter the compression must be by heat or a compressor so waste heat, including that extracted from the contents, has to be disposed of. With a heat pump we specifically want the heat from the compression stage in addition to its normal heat output.

Most materials can be persuaded to give up their heat to a heat pump, in extreme circumstances even if the source is below freezing point. But clearly it is better to concentrate upon the most effective sources. One can weigh up the pros and cons of the most common sources, starting with air. It is free and normally warm enough. It also causes little trouble within the coils. The drawback is the quantity required, as a small heat supply will warm a large volume of air, and so large fans may be needed. River water may possibly be available, and in some areas downstream of an industrial site may be warmer than the norm. The flow should be enough for consistent operation. In this case the snags are possible pollution or corrosive contamination causing trouble in the system. It must not be cooled below freezing point, but a thermostatic control switch can look after this. Well water normally stays above freezing point but the supply may be too limited for continuous operation.

Below the surface, earth is normally above freezing point even in winter and coils can be sunk in the ground. Even if the ground is slightly frozen around the coils, the system should still be effective. Possible corrosion of the coil material must be kept in mind, and there is the problem that after a large withdrawal of heat the ground surrounding the coils will take a while to warm up again.

If there is no available source of low-grade heat, solar panels can be used. As the heat from a solar panel can vary with the sunshine, the idea of using a heat pump in conjunction with solar heat is promising. However, this is not as yet financially attractive for domestic use since heat pumps are expensive and may at the moment be thought of more as gilding the lily if an efficient solar heating system is installed. In poor solar conditions the heat pump may have an efficiency of only 2.0 or even a little less. That is, for every unit of energy put into driving the system only two will be gained. This is worth having but probably not at current costs of equipment. In better conditions the efficiency may be as much as 8.0, but at this stage the domestic user is getting enough from his solar system unaided.

'Heating' so far has usually referred to hot water systems, essentially for space heating via radiators, but has excluded the possibility of temperatures high enough for cooking. Here the addition of a heat pump with an exchanger withstanding temperatures well above boiling point, without boiling itself, offers interesting possibilities; but such equipment—anyway for domestic use—is not yet available.

8
Central Heating

Among the *cognoscenti* of energy conservationists central heating, once a daydream for most of us, has become a pair of dirty words. That is central heating involving heating and pumping hot water to heat up pipes and metal radiators before one BTU of heat enters appropriate rooms. With a well insulated house and one heat source the controlled movement of warm air should be all that is required for space heating. Water heating should be confined to the hot tank for domestic hot water only, probably backed up with solar assistance. Even when solar heating makes the water only lukewarm, this is much cheaper to heat than water straight from the cold tank or mains, and the extra heat can be provided from the same heat source as that used for space heating. The only extra heat energy required is for cooking and even this may be unnecessary if the heat source itself is incorporated in the cooker.

Those building from new should give attention not only to the most familiar forms of insulation but to floors. At the same time they should consider ducting of warm air to all the rooms from the main point of heat output. It is at this stage that ducting is most easily incorporated into the overall design of the house. In the early days of car heating a small radiator under the dashboard, piped from the hottest part of the engine and incorporating a fan, was standard equipment; it was aptly called a 'fug-stirrer'. With good ducted heating today the recycled warm air is filtered and at the same time fresh air is taken in from the exterior. Thus a balance is struck between economy and health. Full air conditioning, which includes cooling the air when ambient temperatures are uncomfortably hot, consumes a good deal of energy in its operation unless solar cooling is used (just as heat is used in refrigeration). But in a temperate climate this adds unnecessarily to cost and energy usage.

Let us take the example of the commonest family housing—a semi-detached (duplex) house with back-to-back kitchens, and the boiler for water heating in the kitchen. If the decision is to install central heating, the normal procedure is to replace the old heating unit with an automatic one, most often with an automatically controlled supply of gas or oil, or possibly solid fuel. Then the plumbing starts with narrow bore (expensive) copper pipes to all rooms to supply newly supplied metal radiators. The cost, even spread over some years, is considerable. Yet the kitchen is one of the least central rooms in the house and the only reason the heater is placed there is because that is where the boiler always has been. Unimaginative thinking is blocking the chance for a new approach—and the kitchen boiler space can be used for far more practical kitchen use.

Such a typical house has a fireplace in the living room, incorporating a chimney breast with a recess to either side, possibly used for bookshelves. The alcove nearest the centre of the house is a suitable place for a heating unit. The unit can easily be hidden from view and so will be cheaper because there is no need for expensive stove enamelled cladding for appearances sake, as would be necessary in the kitchen.

The system has several things in its favour. It is ideal for the DIY man to install. It costs a great deal less than conventional systems with radiators. Running costs are less. And although access to the heater must be provided for service purposes, the system is out of sight, hidden behind the décor of the living room. Also, nearly all the work takes place in one room, cutting labour time and minimising mess. The only possibly adverse factor is the loss of up to six inches in the ceiling height of the living room.

The Warmheart system, as it is called by its designers (Combined Energy Consultants Ltd, Llanelli, South Wales), provides domestic hot water but, in common with heating systems in most parts of the western world, provides heating by means of warm air rather than water (Fig 42). Warm air itself has advantages in addition to cutting out expensive plumbing and the costly transfer of heat-to-water-to-metal-to-air for each room. As the movement of air is direct from heater to points of need, it is very quick. For

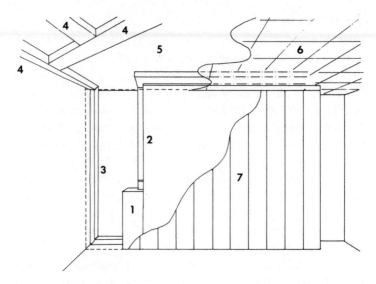

Fig 42. The Warmheart system: (1) Warm air (and water) heater; (2) Air intake; (3) Main duct; (4) Secondary ducts; (5) Existing ceiling; (6) Acoustic tiled ceiling; (7) Facia.

example, if you go out for even a few hours you would probably leave at least some radiators on in an orthodox system, for if rooms get cold it takes time for the water/radiator systems actually to warm up the rooms to normal set temperature. But with air you can turn off the supply when you go out, then when you turn on again you have immediate warm air where you want it. The additional cost savings here are obvious. Also wall spaces are clear of radiators, making rooms easier to keep clean and to redecorate. A fan can be sited out of earshot but its effect is to provide warmth immediately through the grilles. As the air is filtered, it is dust-free, so that the room tends to keep cleaner as well as being more healthy.

The system is based on gas as the fuel and is approved by the Gas Board in the UK. It can also be programmed. The decorative aspect requires covering not only for the heater but also for the fireplace, using the same veneer or whatever you choose. While there must be provision for complete access to the stove occasionally, the controls can be incorporated in a small, unobtrusive opening section so that they can be set or reset at any time.

Once the heater unit is fitted in the chimney breast alcove a duct is used to ceiling level and then subdivided into further ducts to distribute heat to the remainder of the house (Fig 43). Each room has an inconspicuous grille, most often in the floor (upstairs) or in the wall (upstairs or downstairs). These grilles are manually adjusted between fully open or shut—more easily than using the knobs on radiators and indeed with quick visual observation of the setting.

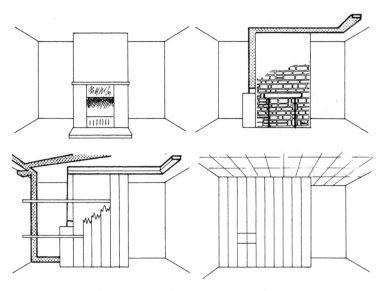

Fig 43. (*top left*) A typical living room in a semi-detached house; (*top right*) The chimney blocked in, heater unit and one duct installed; (*lower left*) The living room ducting completed, with stays set for the acoustic ceiling tiles and for panelling in the heater, allowing an access door for service; (*lower right*) Job complete, with only the controls visible.

In the Warmheart system the ductwork is of Nilflam, which is self-insulating and as light as polystyrene. The appeal to the DIY enthusiast (and it also cuts cost if a contractor is employed) is the simplicity on site of forming the ducts from Nilflam 'sheet' and installing them on existing surfaces. The illustration (Fig 44) shows that with the special knife provided the Nilflam can be chamfered to give a rectangular section using sealing tape on the final joint. The usual ducting layout is shown in Fig 45.

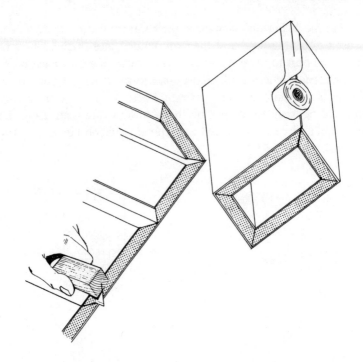

Fig 44. The Nilflam ducting is cut with a special knife, folded into shape, and sealed with broad adhesive tape.

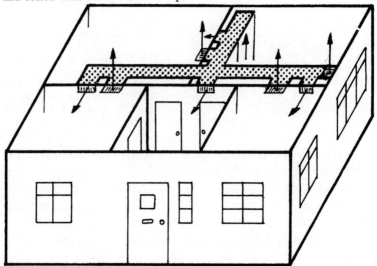

Fig 45. Schematic plan of the warm air ducted layout.

The loss of ceiling height in the living room has been mentioned because nearly all ducting is on this one ceiling, and the ducts are each about four inches in depth. This calls for a false ceiling, which is made of acoustic tiles and may well add to the appearance of the room. Space does not permit details of installation, but it is much easier than one might suppose, and everything required can be supplied in the 'package kit' or from individual suppliers.

Other forms of central heating are commonplace, but beware of an advertising campaign not long ago that provided a living room heater and grilles fitted in doors and walls so that the heat would warm the whole house by convection. To do this the living room would have to be like a boilerhouse!

If you do select an orthodox radiator system, then material such as cooking foil or polished aluminium sheet can be used on the wall behind each radiator to reflect heat. Shelves above radiators mounted against exterior walls also help to deflect warm air into the rooms (Fig 46).

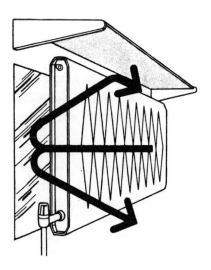

Fig 46. Reflecting radiated heat, particularly from outside walls. The shelf also throws heat into the room.

9
Energy in Industry

While this book is intended to help and advise the occupants of private dwellings, many readers will have some connection with industry. Some reference to industrial husbandry of energy (in the form of cutting costs) is therefore worth a note. In the UK, the Department of Energy (DoE) has several schemes available—some free. For example, there is the Energy Survey Scheme (ESS). This is intended to encourage industrial, commercial and public sector organisations to employ consultants, with the expectation that energy consumption will be reduced and, of course, costs cut.

Big companies will already know of such matters because every aspect of manufacturing processes is under the magnifying glass. But smaller companies may be missing out. The scheme offers up to £75 towards the cost of a one-day consultant's visit and up to half the cost of an extended survey. A list of suitable consultants may be obtained from the Energy Conservation Department of the DoE, and from various other sources listed in a leaflet (ESS5) from the DoE called 'Energy Survey Scheme'. The DoE main headquarters are at Thames House South, Millbank, London SW1P 4QJ. The application form for a repayment is a tear-off slip in the leaflet.

Another leaflet from the DoE covers extended surveys (Form ESS/F1). This is for an 'energy audit' for a detailed and comprehensive review of overall energy usage. Normally half the cost will be met by the DoE provided that approval in writing is obtained first. Again this leaflet covers various useful addresses including Wales and Scotland, and the DoE's Small Firms Information Centres. Again from the DoE is a booklet called 'Energy Saving in Industry', which gives a broad outline of how to set about energy saving. There is also a guide to financial grants for demonstrating new ways of saving energy in industry, commerce, transport, buildings, agriculture, horticulture, and so on.

Finally there are 15 booklets available from the same source:

No.	Title
1 & 11	Energy Audits, Parts 1 & 2
2	The Sensible Use of Latent Heat
3	Utilisation of Steam for Process and Heating
4	Compressed Air and Energy Use
5	Steam Costs and Fuel Savings
6	Flash Steam and Vapour Recovery
7	Degree Days
8	The Economic Thickness of Insulation for Hot Pipes
9	How to Make the Best Use of Condensate
10	Controls and Energy Savings
12	Energy Management and Good Lighting Practices
13	The Recovery of Waste Heat from Industrial Processes
14	Economic Use of Oil-fired Boiler Plant
15	Economic Use of Gas-fired Boiler Plant

Clearly, if you are a member of a small to medium-size firm, it will be worth a stamp or a telephone call for appropriate publications. Equally, as a small manufacturer you will know bureaucracy machinations. It took the author three long distance calls to the DoE Press Office—each time with urgency the keynote—to get the material just described a month later.

10
Planning Permission

'Planning permission' is often thought in the UK to be entirely negative: whatever you want to do, don't—at least until you have Authority's written approval. The staff of Planning Departments tend to be considered an interfering nuisance, but this simply is not so. Just the example quoted in Chapter 1 (pp 11–12) may prove this point. There are many 'wrinkles' of this sort, apart from more obvious considerations such as the security and peace of mind of the neighbours. So it is always best, even before putting anything on paper, to have a talk with the local planning people to find out not only how they react to your scheme but also to get what may be most helpful suggestions. Everyone is better and more cooperative when in the euphoria of giving advice!

A point to watch is that when taking heat or other forms of energy from your environment you must consider the possible effects upon other people. Waterpower is a good example. Taking out heat with a heat pump will reduce water temperature, making it perhaps more prone to freezing, and a slowing of the water flow could cause silting elsewhere. Possibly plant and fish life could be affected, thus reducing amenities for those downstream. With a windmill, the two considerations are safety and appearance. There must be no possibility of a blade flying off in a gale and endangering neighbours, leave alone the whole mill and pylon falling down. The mill will also have to be capable of being sited without disfiguring the surroundings. And no structure must rob near neighbours of their existing light. It is not a bad idea to discuss plans directly with neighbours, for you may uncover a mutual need resulting in a more communal approach to the use of alternative energy sources.

Mainly, planning authorities need to be informed about anything that will alter the use, construction or appearance of your house (excepting decoration). One of the most difficult permissions to obtain in an urban area is for a windmill, as there is the possibility of danger and also a change in the skyline of the area. You will also need permission for solar panels on the roof as they change the construction, use and appearance, but this should be easier. If the planning people are difficult remember that although all buildings come under their control, constructions separate from the house may be more freely erected. If hard pressed by the authority about solar panels, you have the possibility to site these unattached to the house, carrying the hot water in lagged pipes. Storage tanks for hot water or accommodation for storage batteries can be housed separately from the house without need of planning permission, but it is still as well to let the authority know what you are doing.

After a preliminary talk with the planning department you will then need to provide drawings of your detailed intentions. At this stage there is no need for the type of drawings which only an architect or draughtsman can produce. It will be enough to produce amateur drawings, to scale if possible, with constructional details written in as necessary. There are three basic drawings, elevation, plan and section. If you can manage it, possibly with the help of friends, add a perspective to show what the finished job will look like. If one takes an example, like solar panels on the roof, then the elevation will show the appearance. The section would be drawn through the solar panel(s) and roof to show the materials used and the security of the mountings. The plan view would show the proportional amount of roof taken by the panels. The perspective, if possible, will show the general appearance related to the surroundings.

Much of this concerns your own DIY plans, but if you are using commercially made units, the literature from the company concerned will be handy. While the planning department will be primarily concerned with safety and appearance, it is as well to show the performance figures you have calculated. With some notable exceptions (like the new town of Milton Keynes), local authorities are often well behind the times on alternative energy and allied subjects. The extra information you give them may help them to bone up on the subject, so that they will be in a position to help others in their search for a more sensible approach to the use and conservation of energy.

After discussions with the planning department, start on finished drawings to scale. Again, you may be able to do these yourself or, at the other extreme, need an architect. In any event, include suggestions made by the planners, if necessary writing in details of some aspects of the construction work. Get a copy of the building regulations that apply in your area. As it normally takes two months or more to get approval or refusal, it is as well to try to get every detail of regulations right rather than have to go back to the end of the queue with your suitably modified application. Once you have permission, the local building inspector will be keeping an eye on the work as it progresses. Take care to keep him constantly informed, as he can legally tell you to dismantle anything which—even to a trivial degree—conflicts with either the regulations or the details of the approved constructional work.

Financial aid

Building societies will usually help with an increased mortgage, because the chances are that you are increasing the resale value of your house. If you have a mortgage, the society will in any case have to be informed about constructional changes. Its basic interest in your work is to ensure that the house remains at least as saleable as before. What it thinks to be unsightly gimmicks may be turned down. However, the chances are if the building society will not help with funds, outside contractors for some major items may have an extended payment scheme.

Like the planning authority, your building society will need details of what you have in mind, but photocopies of whatever has been sent for planning approval should cover this. Generally, alternative technology is acceptable provided that conventional systems are not discarded in over-enthusiasm. Building society managers' attitudes are predictable; they want to protect their investment. So solar heating as a back-up is usually acceptable but they prefer proprietory systems to DIY. Wind as a standby may be satisfactory but safety and aesthetic factors may militate against it in an urban area. Methane, unlikely to be favoured. Heat pumps, satisfactory at your expense but not necessarily qualifying for mortgage increase to cover cost. Waterpower, no problem and could obviously increase house value.

Finally, check with current availability of government financial help for energy conservation schemes. At the time of writing a grant is available in the UK for insulation schemes valued at 66 per cent or £50, whichever is the smaller.

Appendix I: Advice on Insulation

Department of Energy list, giving local sources first, then associations or federations.

- your local Domestic Coal, Electricity or Gas Consumers Council.
- your coal merchant, solid fuel or oil supplier, electricity or gas showroom.
- your local office of the Solid Fuel Advisory Service or Living Fire Centre (call Sunderland 73578 for the address of your nearest contact).
- builders' merchants and authorised home improvement centres.
- Home Heating Enquiry Line:
  Heating and Ventilating Contractors Association, 34 Palace Court, London W2 4JG. Tel: 01-229 5543. (For names of contractors to provide impartial advice on heating systems.)
- Building Centres in Birmingham, Bristol, Cambridge, Glasgow, Liverpool, London, Manchester, Nottingham and Southampton (for information on materials, controls, appliances and suppliers).

(See your local telephone directory for addresses and telephone numbers as necessary.)

Loft insulation

Eurisol UK (Association of British Manufacturers of Mineral Insulating Fibres), 64 Wilton Road, London SW1V 1DE.
Tel: 01-828 0151.
National Association of Loft Insulation Contractors (NALIC), 178-202 Great Portland Street, London W1N 6AQ.
Tel: 01-637 7481.

Cavity wall insulation

Agrément Board, P.O. Box 195, Bucknalls Lane, Garston, Watford WD2 7NG. Tel: Garston (Herts) 70844.

National Cavity Insulation Association, 178-202 Great Portland Street, London W1N 6AQ. Tel: 01-637 7481.

Double glazing

Glass & Glazing Federation, 6 Mount Row, London W1Y 6DY. Tel: 01-629 8334.

Controls

The Hevac Association—Automatic Controls Group, Unit 3, Phoenix House, Phoenix Way, Heston, Middlesex TW5 9ND. Tel: 01-897 2848/9.

Hot water cylinder jackets

Insulating Jacket Manufacturers Federation, Little Burton West, Derby Street, Burton-on-Trent, Staffs DE14 1PT. Tel: Burton-on-Trent 63815.

Draughtproofing

Draught Proofing Advisory Service, P.O. Box 305, Bushey, Herts WD2 3HF. Tel: 01-950 5310

Appendix II: Solar Energy Companies

This list has been compiled by the Solar Energy Unit of Cardiff University. It must be made clear that the Unit does not claim that it is exhaustive in spite of its being seemingly comprehensive. The Unit is at pains to make clear that the companies have not been checked for technical competence or quality of product. However, the list is useful because it is broken down into the various categories which the reader would want. The key is as follows:

Code Letter	Meaning
M	— Manufacturer or importer of solar absorbers unless otherwise stated
D	— Distributor
I	— Installer
A	— Producer of ancillary equipment, ie equipment other than the panel
F	— Research and development
Hp	— Heat pumps
Sp	— Swimming pool
C	— Consultancy (NB this designation is self-styled and given without prejudice)
K	— DIY kits

Acoustics & Environmetrics Ltd
Ruxley Towers
Claygate
Esher, Surrey KT10 0UF
Tel: 0372 67281/3
Code D

Air Distribution Equipment
(M & W) Ltd
64 Whitebarn Road
Llanishen
Cardiff CF4 5HB
Tel: 0222 40404
Code M, I, A, F, C

Alcan Windows Ltd
Winterstoke Road
Weston Super Mare, Avon
Tel: 0934 27511
Code D, A, F

Alfa Joule Ltd
38 South Avenue
Chellaston, Derby DE7 1RS
Tel: 0332 701127
Code I, A

Alcoa (GB) Ltd Distribution
Centre
PO Box 15
Droitwich
Worcs WR9 7BG
Tel: 090 57 3411
Code M, D

Allied Windows (South Wales) Ltd
Gulf Works
Penarth Road
Cardiff CF1 1YS
Tel: 0222 397788
Code I

Aluglaze
Alcan Booth
Raans Road
Amersham, Bucks
Tel: 02403 21262
Code D

Aluglaze Home Improvements
Cape House
787 Commercial Road
London E15
Code D

W. E. Anfield & Co Ltd
Station Road
Chester CH1 3PA
Tel: 0244 312404
Code D, I, C

Antarim Ltd
Solar Works
New Street
Petworth, Sussex GU28 0AS
Tel: 0798 43005
Code M, D, A, Hp

Appliance Components Ltd
Cornwallis Street
Maidenhead, Berks SL6 7BQ
Tel: 0628 32323
Code A (Valves)

Aquasol (UK) Ltd
93 King Street
Maidstone, Kent
Tel: 0622 57569
Code I

W. Arundel Ltd
17 Beech Close
Sproatley
Hull, North Humberside
Tel: 0482 812827
Code I, C

Asahi Trading Co Ltd
Asahi House
Church Road, Port Erin
Isle of Man
Tel: 062483 3379
Code M, D

Atkins Research & Development
Woodcote Grove
Ashley Road
Epsom, Surrey
Tel: Epsom 26140
Code C, F

Avica Equipment Ltd
Mark Road
Hemel Hempstead, Herts
Tel: 0442 64711
Code F

Avonray Ltd
61 Rabbit Row
London W8 4DX
Tel: 01 229 0383
Code A (Sp cover)

Aztec Solar Ltd
46 The Vineyard
Richmond, Surrey
Tel: 01 940 6867
Code M, D, I, A, Sp, C, K

111

C. T. Baker Ltd
Market Place
Holt
Norfolk NR25 6BW
Tel: 026 371 2244/5
Code I

Barum Solarheat
14 Greig Drive
Goodleigh Rise
Barnstaple, Devon
Tel: 0271 75962/3377
Code I

Bath Engineering Ltd
Wood Street, Lower Bristol Road
Bath, Avon BA2 3BJ
Tel: 0225 318976
Code M & I

Beau Design Services Ltd
Pipeline House
25-30 Theobald Street
Borehamwood, Herts WD6 4SG
Tel: 01 953 4065/4423
Code A (Storage slabs)

Bexley Glass Ltd
37 High Street
Bexley, Kent DA5 1AB
Tel: 029 53311
Code M, D, I, A

Bracaster Ltd
25 East Street
Blandford Forum
Dorset DT11 7DU
Tel: 0258 54380
Code M

S. Briggs & Co Ltd
New Street
Burton-on-Trent
Staffs DE14 3QL
Tel: 0283 68232
Code M

Brinsea Solar Engineering Ltd
West Brinsea
Congresbury, Avon
Tel: 0934 852082
Code F, C

British Industrial Plastics Ltd
Aldridge Road Streetly
Sutton Coldfield
West Midlands B74 2DZ
Tel: 021 353 2411
Code A (Plastic covers)

Building Heat Conservation
Eyot House
Shoreham
Nr. Sevenoaks, Kent
Tel: 095 92 2801
Code M, I, Sp, K

Calor Sol Ltd
(Subsid. of A. T. Marston & Co Ltd)
Lancaster Road
Shrewsbury SY1 3NG
Tel: 0743 51578
Code M, I, K

Complete Gardeners Ltd & Cotswold Pools
Highnam Court
Gloucester
Tel: 0452 23078/22868
Code D, I, Sp, K

Concorde Ecru
154 Norwich Road
Dereham, Norfolk
Tel: 0362 4915/4055
Code F

Construction Aids Ltd
9-19 St George Street
Norwich NR3 1AB
Tel: 0603 615407
Code: M, D, I, Sp, Hp, C, A

Coolag Ltd
PO Box 3
Charlestown
Glossop, Derbyshire SK13 8LE
Tel: 04574 322779
Code A (Insulated board)

Corbridge Services
c/o National Chalet Centre
Ullswater Road
Penrith, Cumbria
Tel: Penrith 5881
Code M, D

J. Arthur Cowley Ltd
Lansdowne Road
Cardiff, South Glam.
Tel: 0222 20580
Code I

W. J. Creemer Ltd
112 North Road
Cardiff, South Glam.
Tel: 0222 387426
Code D, I

Dewraswitch Ltd
2 Pit Hey Place
Pimbo
Skelmersdale, Lancs WN8 9PG
Tel: 0695 24270
Code A

Distrimex Ltd
88 The Avenue
London NW6 7NN
Tel: 01 459 1391
Code M, D, A, Hp, Sp, Solar cells

Domshelcon Ltd
Hebden, South Drive
Littleton
Winchester, Hants SO22 6PY
Tel: 0962 880521
Code A, F, C, K

Don Engineering (South West) Ltd
Wellington Trading Estate
Wellington, Somerset
Tel: 082347 3181
Code M, A, F, Sp

Doulton Heating Systems Ltd
School Lane
Knowsley
Prescot, Merseyside L34 9HJ
Tel: 051 546 8225/6/7
Code M, Sp, I

Drake & Fletcher Ltd
Parkwood
Sutton Road
Maidstone, Kent ME15 9NW
Tel: 0622 55531
Code M, Sp

Dunlop Ltd
G. R. G. Division
Cambridge Street
Manchester M60 1PD
Tel: 061 236 2131
Code A

J. Thomas Edwards & Sons Ltd
Unit 1, Vulcan Trading Estate
Leamore Lane
Walsall
Tel: 0922 406514
Code M, Hp, Sp

Electra Air Conditioning Services Ltd
66-68 George Street
London W1H 5RG
Tel: 01 487 5606
Code M

Electricare
38 Dillotford Avenue
Coventry CV3 5DR
Tel: 0203 414064
Code A

Elsy Gibbons Ltd
Simonside
South Shields
Tyne & Wear NE34 9PE
Tel: 0632 561474
Code A (tanks)

The Energy People
75 Stoddens Road
Burnham-on-Sea
Somerset TA8 2DB
Tel: 0278 784298
Code C, D, I, Sp, Hp

Eumed Ltd (Solartherm
International)
Precision House
20 Bridge End
Leeds LS1 4DJ
Tel: 0532 34256
Code M, I

Ferranti Ltd
Electronic Components Division
Gem Mill
Chadderton
Oldham, Lancs OL9 8NP
Tel: 061 6240515
Code M (Solar cells)

Fortress Engineering Ltd
Rockbourne Mews
56a Rockbourne Road
Forest Hill
London SE23 2DE
Tel: 01 699 0457
Code M

Four-T Engineering Ltd
North Dock
Llanelli, Dyfed
Tel: 05542 56041
Code M, D, I, A, F, Sp, K

D. F. Furnell Plant Services Ltd
Ebberns House
68 Ebberns Road
Hemel Hempstead, Herts
Tel: 0442 55244/55848
Code D, I, A, C, K

G & S Solar Power Systems
(Bradford) Ltd
Crown Works
Birkshall Lane
Bradford BD4 8TD
Tel: 0274 21045
Code M, D, I, A, F, Sp, K

General Technology Systems
162D High Street
Hounslow
Middlesex TW3 1AA
Tel: 01 568 5871
Code C

Granomead Ltd
Unit 27, Huffwood Trading Estate
Brookers Road
Billingshurst
Sussex
Tel: 0403 752726
Code I, A, C

Greenway Engineering Ltd
1 Greenway Park
Chippenham, Wilts
Tel: 0249 2434
Code A

Grumman International Inc
64-65 Grosvenor Street
London W1X DB
Tel: 01 629 3847
Code M, A, F, (Solar cells)

Guardian Window Co Ltd
Hansa Road
Hardwick Industrial Estate
King's Lynn
Norfolk
Tel: 0553 3180/5202
Code M, D, I, C, F, Sp

Heatguard Solar Systems
Chequers Road
Derby DE2 6AU
Tel: 0332 363748
Code M

J. A. Heating Systems Ltd
PO Box 19
Beaconsfield, Bucks
Tel: 049 46 5255
Code M

Heatmaster Boiler (Newhaven) Ltd
Unit 5, Chaucer Industrial Estate
Duttons Road
Polegate
Sussex BN126 6TF
Tel: 03213 5226
Code M

Heliotechnic Ass. International
5 Dryden Street
Covent Garden
London WC2E 9NW
Tel: 01 240 2430
Code C

Helix Multi Professional Services
Mortimer Hill
Mortimer
Reading, Berks
Tel: 0734 333 070
Code F, C

Hereford Solar Heating
Services Ltd
Little Hill
Orcop, Herefordshire
Tel: 098 14 656
Code D, Hp, M, I, A

Heywood Heating Services Ltd
16 The Halve
Trowbridge, Wilts
Tel: 022 14 3803
Code I

ICI Ltd
Plastics Division
PO Box 6
Bessemer Road
Welwyn Garden City
Herts AL7 1HD
Tel: 070 73 23400
Code M

IDC Consultants Ltd
Stratford upon Avon
Warwickshire CV37 9NJ
Tel: 0789 4288
Code I, C, F

IMI Range Ltd
PO Box 1
Stalybridge
Cheshire SK15 1PQ
Tel: 061 338 3353
Code A (Storage tanks)

Inco Europe Ltd
R & D Centre
Wiggin Street
Birmingham B16 0AJ
Tel: 021 454 4871
Code A

Industrial (Anti Corrosion)
Services
Britannica House
214-224 High Street
Waltham Cross, Herts
Tel: 0992 28355
Code A (Propylene glycol)

Insolar Technology
Telefield (Bristol) Ltd
Old Shaftesbury Crusade
Kingsland Road
St Phillips
Bristol 2
Tel: 0272 552834
Code M, I, F, Sp

Instrumentation & Control Services
(Hereford) Ltd
Whitehorse Square
Hereford HR4 0HD
Tel: 0432 55341
Code A

Island Environmental Designs Ltd
4 Cedar Hill
Carisbrooke
Isle of Wight
Tel: 098 381 3535
Code M, C, F

JA Heating Systems Ltd
PO Box 19
Beaconsfield, Bucks
Tel: 049 46 5255
Code D, A, F, Hp, Sp, C

F. J. Jones, Heating Engineers Ltd
30 Gorsty Hill Road
Rowley Regis
Warley, Worcs B65 0HD
Tel: 021 559 3658
Code M, I, Sp

KB Commercial Solar Energy Ltd
1B Woodthorpe Road
Nottingham NG3 5QH
Tel: 0602 622847
Code M, I, A, Sp

Kennedys Sunsaver Solar Systems
2 Wellington Road
Bournemouth, Dorset
Tel: 0202 22100
Code I, F, C

Kent Coast Refrigeration &
Engineering Ltd
Henwood Industrial Estate
Ashford, Kent
Tel: 0233 31951
Code D, I, Sp

Kleen Line
School Lane
Knowsley
Prescot
Merseyside L34 9HJ
Tel: 051 227 2124
Code M

Kroneck & Jones Ubbley Wood Ltd
Mechanical Service Engineers
39A George Street
Weston-Super-Mare
Avon
Tel: 33451
Code I, Hp

John Laing Research &
Development Ltd
Manor Way
Boreham Wood, Herts WD6 1LN
Tel: 01 953 6144
Code F, C

Landis & Gyr Ltd
(Confort Control Group)
Victoria Road
North Acton
London W3 6X5
Tel: 01 993 0611
Code A

Percy Lane Group Ltd
83 Colmore Row
Birmingham B3 2AP
Tel: 021 236 3911
Code M

Laurie & Stuart Engineering Ltd
Silver Trees
Waterhouse Lane
Kingswood
Surrey KT20 6HU
Tel: 073 783 2091
Code I, M, Sp

A & D Lee Co Ltd
Unit 19 Marlissa Drive
Midland Oak Trading Estate
Lythalls Lane
Coventry CV6 6HQ
Tel: 0203 664664
Code M

Lennox Industries Ltd
PO Box 43
Lister Road
Basingstoke
Hampshire RG22 4AR
Tel: 0256 61261
Code M

London & Lancashire Rubber Co
Ltd
60 Dowsett Road
London N17 9DF
Tel: 01 808 9959
Code M, D

London Solar Engineering
70 Church Road
Upper Norwood
London SE19 2EZ
Tel: 01 771 3000
Code M, I, Sp, K

Longborough Concrete Ltd
Longborough
Moreton-in-Marsh, Glos
Tel: 0451 30140
Code Sp

Lucas Electrical Ltd
(P & S Division)
Great Hampton Street
Birmingham B18 6AH
Tel: 021 236 5050
Code M (Solar cells)

T. G. Lynes & Sons Ltd
35-45 Caledonian Road
London N1 9BT
Tel: 01 837 6455
Code D, C

3M United Kingdom Ltd
3M House
Wigmore Street
London W1A 1ET
Tel: 01 486 5522
Code A (Black paint)

MB Engineering, Tiverton
Poundsland
Silverton, Nr Exeter
Devon EX5 4HD
Tel: 039 286 310
Code M, D, I, Sp

MPD Technology Ltd
Wiggin Street
Birmingham B16 0AJ
Tel: 021 454 0373
Code A (Selective foil)

McKee Solaronics Ltd
12 Queensborough Road
Southminster, Essex
Tel: 0621 772477
Code M, D, C, Sp, A, F

Mastertherm Ltd
Richmond Chambers
Bourne Avenue
The Square
Bournemouth BH2 6DS
Tel: 0202 294201
Code M, D, I, C, F

Morcol Solar Energy
3 Thorpe Road
Harthill
Nr Sheffield
Tel: 0909 771882/730591
Code M, I, K, A (Controllers)

A. J. Murcott
No 1 Factory Unit
Bishops Castle
Salop
Tel: 058 83 348
Code I

Natenco Supplies Ltd
51B New Briggate
Leeds LS2 8JD
Tel: 0532 454255
Code M

Natural Energy Centre
2 York Street
London W1
Tel: 01 486 4186
Code D

Natural Energy (Jersey) Ltd
The Energy Centre
40 Kensington Place
St Helier, Jersey
Tel: 0534 75221
Code M

Natural Energy Systems
143 Easterly Road
Leeds LS8 2RT
Tel: 0532 653323
Code I, Hp, Sp

Newsun
7 Victoria Drive
Rock Ferry
Birkenhead
Merseyside
Tel: 051 645 1913
Code M

Norden Baines Ltd
16 East Street
Shoreham-by-Sea
West Sussex BN4 5ZE
Tel: 07917 3467
Code I

Oceanware Ltd
Alternative Energy Site
Wenlock Edge
Much Wenlock
Salop
Tel: Much Wenlock 0952-727777
Code M, I, F, Hp, Sp, C, K

OCLI Europe
621 London Road
High Wycombe
Bucks
Tel: 0494 36286
Code M

Pendar Technical Associates Ltd
Hamp Industrial Estate
Old Taunton Road
Bridgwater, Somerset TA6 3NT
Tel: 0278 56888
Code F, Hp, C

Photain Controls Ltd
Unit 18, Hangar No 3
The Aerodrome
Ford, Sussex
Tel: 090 64 21531
Code M (Solar cells)

Pilkington Brothers Ltd
Prescot Road
St Helens, Merseyside WA10 3RR
Tel: 0744 28882
Code F, A, C

Plastic Constructions Ltd
Tyseley Industrial Estate
Seeleys Road
Greet
Birmingham B11 2LP
Tel: 021 733 1331
Code A

Plumbing & Home Heating
Services
5 Braes Mead
South Nutfield, Surrey RH1 4JR
Tel: 073 782 2316
Code I

Polar Electronic Developments Ltd
Domville Road
Liverpool, Merseyside L13 4AT
Tel: 051 2206666
Code M, A, C

Precision Electronics
14-16 Cannon Street
Wellingborough, Northants
Tel: 0933 79585
Code F, A

Prematechnik (UK) Ltd
73-79 Rochester Row
London SW1P 1LQ
Tel: 01 834 6013
Code A

Premier Solar Systems Ltd
35 St Leonards Road
Exeter, Devon EX2 4LR
Tel: 0392 51110
Code I, D, Sp, C, K

Quantock Engineering
Development Ltd
Kingston St Mary
Taunton, Somerset TA2 8H2
Tel: 082 345 349
Code D

RDM Sundial Aluminium Ltd
Unit 24, Dawkins Road
Hamworthy
Poole, Dorset BH15 4JY
Tel: 02013 79401
Code D, I

Rapaway Ltd
Eurohurst House
Oakenshaw Road
Shirley
Solihull, West Midlands BG0 4PE
Tel: 021 745 3144
Code M, I, D, Sp, C, K, A

Redpoint Associates Ltd
Cheney Manor
Swindon, Wilts SN2 2PS
Tel: 0793 28440
Code M

Refrigeration Appliances Ltd
(Delta Group)
Haverhill
Suffolk CB9 8PT
Tel: 0440 2653
Code M, Sp

Regency Swimming Pools
36 Compton Road
Chapel Ash
Wolverhampton WV3 9PL
Tel: 0902 27709
Code D, I, Sp, K

Regent Swimming Pools Ltd
The Pool Park
Kennedy's Garden Centre
Floral Mile
Twyford, Berks
Tel: 073522 3912
Code D, I, Sp, C

Robinson Developments Ltd
Robinson House
Winnall Industrial Estate
Winchester SO 23 8LH
Tel: 0962 61777
Code M, Sp

Ruberoid Contracts Ltd
St Mungo Street
Bishopbriggs
Glasgow G63 1QX
Tel: 041 772 1117
Code A

Ruslink Ltd
39A George Street
Weston-Super-Mare
Avon
Tel: 0934 33451
Code M, A, F

Russel-Cowan Properties Ltd
Solar Energy Division
70 Courtfield Gardens
London SW10
Tel: 01 370 4804
Code D

Ryland Pumps Ltd
Bridgewater Road
Broad Heath
Altrincham, Cheshire
Tel: 061 928 6371
Code A

SES Solar Ltd
134 High Street
Deritend
Birmingham B12 0JU
Tel: 021 773 6713
Code M, D, I, A, C, F, Sp, K

Satchwell Control Systems Ltd
PO Box 57
Farnham Road
Slough, Berks SL1 4UH
Tel: 0753 23961
Code A

Scotia Energy Conservationists Ltd
33 Shore Street
Anstruther
Fife, Scotland
Tel: 0333 310448
Code I

Sealed Motor Construction Co Ltd
Bristol Road
Bridgwater, Somerset
Tel: 0278 4366
Code A (Pumps)

Senior Platecoil Ltd
Otterspool Way
Watford Bypass
Watford, Herts
Tel: Watford 26091
Code M

Serac UK Ltd
Sydney House
Queens Road
Haywards Heath
W. Sussex RH16 1EE
Tel: Haywards Heath 50536
Codes M, D, I, A, F, Sp, C

Sile Woodhall Ltd
Rookery Street
Wednesfield
Wolverhampton WV11 1UU
Tel: 0902 737431
Code M

Simsol Solar Heating Ltd
Clarke Street Industrial Estate
Poulton Le Fylde
Blackpool FY6 8JR
Tel: 0253 886871
Code M, I

Sinclaire Air Conditioning Ltd
22 Queen Anne's Gate
Westminster
London SW1H 9AH
Tel: 01 930 5011
Code M, Hp

J. G. Smith Heating Co Ltd
Harpers Road
Newhaven
East Sussex
Tel: Newhaven 4236
Code D, I

Solapak Products
School House
Great Unsworth
Washington
Tyne & Wear NE37 1NU
Tel: 0632 464646
Code D

Solar 2000
Norwich Union House
Lichfield Street
Walsall WS1 1BR
Tel: 0922 614 308/9
Code D, I

Solar Apparatus & Equipment Ltd
Brunel Road
Newton Abbot, Devon
Tel: Newton Abbot 3003
Code M, Sp

Solar Collector Designs Ltd
11 Boyn Hill Avenue
Maidenhead, Berks
Tel: 0628 24909
Code M, D, I, C, F, A, K, Sp

Solar Dynamics Ltd
111-113 High Street
Wandsworth
London SW18 4HY
Tel: 01 874 2729
Code M

Solar Electronics
388/400 Manchester Road
Sudden Rochdale OL11 4PE
Tel: 0706 33438
Code A

Solar Energy Developments
Bay 8
16 South Wharf Road
London WC2 1PF
Tel: 01 402 3203
Code C, F

Solar Energy Industries
4 Fleming Close
London SW10 0AH
Tel: 01 352 2703
Code M, I, C

Solarflair
88 The Avenue
London NW6 7NN
Tel: 01 459 1391
Code F, C

Solar Heat Ltd
99 Middleton Hall Road
Kings Norton
Birmingham B30 1AG
Tel: 021 458 1327
Code M

Solar Heating
931 Wimbourne Road
Moordown
Bournemouth BH9 2BN
Tel: 0202 523726
Code I

Solar Heat Panels
Chattenden Lane
Chattenden
Rochester, Kent
Tel: 0634 270018
Code M, A, F, Hp, Sp, K (Heat storage)

Solar Hydro Electric (GB) Ltd
20-21 Witham
Hull, Humberside
Tel: 0482 23597
Code M, D, I, A, Hp, Sp, C

The Solar Installations Co
27 Lovelace Close
Rainham
Gillingham, Kent ME8 9QN
Tel: 0634 362249/35092
Code I, C, F

Solarpave Company
Dolphin House
Hollybush
Ledbury, Herefordshire
Tel: 053 181 441
Code M, Sp

Solar Plumbing & Heating Co
37 The Chine
Winchmore Hill
London N21 2EE
Tel: 01 360 7433
Code C, P, I, Sp

Solarsense Ltd
48-52 Goldsworth Road
Woking, Surrey GU21 1LE
Tel: Woking 66121
Code M, D, I, F

Solar Technic
11 Glenwood Lane
West Moors
Wimborne, Dorset
Tel: 0202 876646
Code C, D, I, K, Hp, Sp

Solartherm Energy Systems
396 Cheetham Hill Road
Manchester 8
Tel: 061 795 0717
Code M, I, D

Solartraps (Surrey) Ltd
70 East Street
Epsom, Surrey KT17 1HF
Tel: Epsom 23544
Codes D, I, Sp, K, Hp, C

Solar Water Heaters Ltd
153 Sunbridge Road
Bradford, W. Yorks BD1 2PA
Tel: 0274 24664
Code M, D, I, A, Sp, K

Solarway Systems (Scotland) Ltd
13-15 Tower Street
Edinburgh EH6
Tel: 031 554 0358
Code I

Solchauf
Unit 5 Rash's Green
Toftwood
Dereham, Norfolk NR19 1JG
Tel: 0362 2114
Code M, D, I, A

Solight Systems
Solar Heating Engineers
7 The Green
Totley, Sheffield S17 4AT
Tel: 0742 366471
Code D, A, K

Southwest Solar Heating
Parford
Chagford, Devon TW13 8JR
Tel: 064 73 2221
Code I, Sp, C

Space Energy Services (Solar) Ltd
75 Deanston Drive
Shawlands
Glasgow
Tel: 041 649 2857
Code I

Spearhead Energy Conservation Ltd
Shalden
Alton, Hampshire GU34 4DS
Tel: 025683 412
Code D, I, C, Hp, Sp

Spencer Solarise Ltd
North Way
Walworth Industrial Estate
Andover, Hants
Tel: 0264 51625
Code M, D, I, F

Steels Engineering Ltd
Leechmere Works
Sunderland
Tyne & Wear SR2 9TG
Tel: 0783 210813
Code I

Stevenson Heating Ltd
34 West Common Road
Hayes
Bromley, Kent
Tel: 01 462 8822
Code I, Sp, C, K

Stiebel Eltron Ltd
25 Lyveden Road
Brackmills
Northampton NN4 0ED
Tel: 0604 66421
Code M

Stuart Energy
5 Orbel Street
London SW11
Tel: 01 223 1485
Code C

Stuart Gray (Merchants) Ltd
Wylds Road
Bridgewater, Somerset
Tel: 0278 57011
Code M, I, A, Sp, C, K

Sunbryte Home Improvements Ltd
Deanfield Mills
Asquith Avenue
Morley, Leeds
Tel: 0532 534542
Code A, D, I

Sunharvester Ltd
Longridge Trading Estate
Knutsford, Cheshire WA16 7BR
Tel: 0565 52941/54651
Code M, D, I

Sunpower Ltd
Coombe Park
Chillington
Nr. Kingsbridge, S. Devon
Tel: 054853 347
Code C, D, Sp, I, K, Hp

Sunreaper Ltd
Sunreaper House
1 Lowther Gardens
Bournemouth, Dorset
Tel: 0202 302176
Code I

Sunsense Ltd
1 Lincoln Road
Northborough
Peterborough
Tel: 0733 252 672
Code M, D, I, A, F, Sp, C, K

Sussex Solar Services
53 Grove Road
Eastbourne, E. Sussex BN21 4TX
Tel: 0323 20282
Code D, I

Sussex Solar Systems
78 New Church Road
Hove, East Sussex BN3 4FN
Tel: 0273 730920
Code I

TI (Group Services) Ltd
Bridgewater House
Cleveland Row
St James's
London SW1A 1DG
Tel: 01 839 9090
Code F

Technica Solar Energy
88 The Avenue
London NW6 7NN
Tel: 01 459 1391
Code D, I, C, F

Thermal Energy Components Ltd
Park House
20 Bentinck Road
Nottingham NG7 4AD
Tel: 0602 704515
Code M, F, C

Thermoray Ltd
33 High Street
Cowbridge, South Glamorgan
Tel: 04463 4639
Code M, D, I, Sp

Torday Group
West Chirton Industrial Estate
North Shields NE29 8RQ
Tel: 089 45 75577
Code A

Totten Electrical Sales
Magnetic Pump Division
Southampton Road
Cadnam, Hants SO4 2NF
Tel: 042 127 3136
Code A

F. G. Trew & Son Ltd
49-50 King Edward Road
Swansea SA1 4LN
Tel: 0792 56197
Code D

Triad Architects
9 Tufton Street
London SW1
Tel: 01 222 0551
Code C

UF Foam Consultants
184 Cedar Road
Earl Shilton, Leics
Tel: 0455 42965
Code M

Universal Solar Systems
192-4 Stafford Street
Walsall, W. Midlands
Tel: Walsall 22466
Code M, D, I

Verdick Ltd
Colliers Corner
Lane End
Nr High Wycombe
Bucks
Tel: 0494 881 254
Code M, A, Sp

Vulcanheat
107a London Road
Leicester
Tel: 0533 51709
Code D, I, C

WB Solar Economy
Balksbury Hill
Upper Clatford
Nr. Andover, Hants
Tel: 0264 51522
Code M, D, I, A, Sp, F, C, K

WR Heat Pumps Ltd
Unit 2
The Causeway
Maldon, Essex
Tel: 0621 56611
Code Hp, C

Washington Engineering Ltd
Industrial Road
Washington, Tyne & Wear
Tel: 0632 463001
Code M

Wavin Plastics Ltd
PO Box 12
Hayes, Middlesex UB3 1EY
Tel: 01 573 7799
Code A, Sp

Western Automation
44-50 Collingdon Road
Cardiff CF1 5ET
Tel: 0222 396446
Code A (Controllers)

J. Williams (Energy Services) Ltd
The Furlong
Berry Hill Industrial Estate
Droitwich, Worcs WR9 9AJ
Tel: 09057 3701
Code M, I, Sp

Wingrove & Rogers Ltd
Domville Road
Liverpool L13 4AT
Tel: 051 220 4641
Code A

Winson Group
St Anne's House
45 Park Street
Luton, Beds
Tel: Luton 421861
Code A, Hp, M

Index

absorber, flat plate, 32-3, 49-52, 59
aerofoils, 74, 77
Agrément Board, 11, 109
air conditioning, 96; solar, 28, 40
air, warm, 6-7, 96, 97-100
alternators, 72, 89
aluminium, 33, 34, 101
America, North, 77; South, 64
anemometer, 73
antifreeze, 37, 39, 42
apple wood, 67
architects, 9, 105, 106
ash wood, 66
Australia, 40

back boiler, 69
Banki, Donat, 88; — turbine, 88
baths, bathing, 6, 7, 18, 24, 26, 59
batteries, storage, 72, 78, 79, 89, 105
'bio-gas', 91
bobweights, 74-5
Britain, 72
British Standards Institution, 11; — Specification, 22
building societies, 106 see also mortgages

Canada, 77, 78
carbon monoxide, 68
Cardiff University Solar Energy Unit, 110
casing, for solar collectors, 32-4, 49-54
Catenary mill, 77
cavity walls see insulation
central heating, 6, 12, 24, 29, 35, 96-101
Centre for Alternative Technology. Machynlleth, 4, 77, 92
chainsaws, 65-6
charcoal, 68
chimneys, 6, 66, 69-71 see also stovepipes
chipboard, 15, 50
cladding, 67, 71, 97
clips, 43, 45
coal, 60, 64, 67, 94

collectors see solar heating
Combined Energy Consultants Ltd, 69, 78, 97
combustion stoves, 66-71 see also individual headings
compression coils, 94-5
concentrators, 27-8
condensation, 14, 18, 20
controls, 23-5, 35, 37, 43, 109 see also programming; thermostats; time switches
cooling, 28, 40, 96
cooking, 95, 96
copper, 33, 39, 40, 54-8
corrosion, 14, 34, 39, 43, 95
curtains, 7, 10, 20

DAF model mill, 77
dams, 81, 82-4
Darrieus, 77; mill, 77
deepfreeze, 7, 9
Denmark, 72, 78; State Electricity Board, 74
descaling, 54
digesters, methane, 91-2
DIY, 16, 19, 20, 26, 29, 33, 34, 40, 43, 47-59, 78, 85, 88, 89, 97, 99, 105, 106
doors, 18-19
Doughnut stove, 67-8
draught-proofing, 7, 17-19, 109
driftwood, 67

ecology, 59, 64
electric fire, 9, 61; wiring, 14-15
electricity, 60, 72, 77, 79, 87, 89; — Board, 15
elm wood, 64, 66
Energy, UK Department of, 11, 19, 22, 25, 102, 108; Conservation Department, 102; Survey Scheme, 102
Energy Primer, 89
ethylene-glycol, 37
exchangers, heat, 37, 39-40, 42, 54-9, 95

125

fans, 37, 38, 40, 94, 96, 98; extractor, 18
Faraday, 40
fertilizers, 64, 90, 91, 92
Finland, 76
floors, 10, 21, 96
foam, 11, 18
Forest Service, US, 64
Forestry Commission, UK, 65
Franklin, Benjamin, 68; stove, 68-9
frost protection, 17, 23, 37, 41

gas, 17, 60, 90-1, 97, 98, bottled, 17, 18; — Board, 98
gearing, 81, 86, 89
generators, 72-5 *passim*, 77, 80, 81, 86, 89
Germany, 88
glass, 31-3, 48-54; fibre, 13, 15, 17, 39
glazing, double, 7, 10, 19-21, 33, 109; treble, 7, 10, 21
grants, 13, 27, 102, 107
granular loosefill, 13, 15-17, 22
greenhouses, 38; 'effect', 28, 31, 33, 38
guarantees, 11

hardboard, 15-16, 22
hardwood, 50, 65, 67
heat loss, 5-7, 9-10, 21, 24
heaters, immersion, 5, 21, 22, 24, 40, 41, 79; night storage, 21, 37-8; solid fuel, 63-4, 67-71
hickory wood, 66
Holland, 72
hot water systems, 21-5, 39, 37-43, 96; two-tank, 41-3
hydrochloric acid, 54

industry, 102-3; 'Energy Saving in', 102
insulation, 7, 9-25, 32, 37, 38, 49, 51, 52, 60, 64, 96, 107, 108-9; floor, 21, 96; hot water systems, 21-3; loft, 12-17, 108; walls, 7, 10-12, 109

jackets, insulating, 22, 109
Japan, 68

'Kitemark', 22

lagging, 5-6, 14, 21, 23
local authorities, 11-12, 13, 19, 27, 90, 104-7
lofts *see* insulation
'luffing', 75

Magnus principle, 76
matting, 13, 17
McCartney, Kevin, 62
McLaughlin, T.P., 90
metal strip, 18-19
methane, 90-2, 106
methanol, 37
Michell, A.G.M., 88; — turbine, 88
mortgages, 61, 106, 107
mounting, solar, 45-7; windmill, 73

New Mexico, 38
Nilflam, 99
North Sea, 90
nuclear energy, 60

oak wood, 66
oil, 9, 17, 18, 97

painting, 33, 52
paper, waste, 67, 91
paraffin, 17
Pelton wheel, 88-9
Pembroke heater, 69
pine wood, 66
pipes, 5, 13, 15, 17, 21-2, 39, 40, 54-7, 97 *see also* plumbing
planning permission, 12, 19, 104-7
plastic, 15, 16, 18, 21, 33
plugs, 48, 52
plumbing, 6, 14, 43, 46, 54, 97
pollution, 27, 59, 64
Poncelet wheel, 85-6
Portola Institute, 89
power stations, 93-4
Prism Press, 89
programming, 23-5, 98
pumps, 34-6, 43, 56, 86; heat, 93-5, 104, 107
Pyrenees, 27

radiation, 30-5, 37
radiators, 6, 26, 33, 39, 47-9, 52, 54-5, 95-8 *passim*
Reading University, 77, 78
refrigeration, 28, 40, 96
refrigerators, 9, 28, 39-40, 93, 94

safety, 67, 73, 75, 78, 80, 104, 105, 106
Savonius, Sigurd, 76; — rotor, 76, 77
sealing, 34, 52, 53, 54
sensors, 36, 43, 69
sewage, 90-1
showers, 7

Small Firms Information Centre, 102
sliprings, 75
solar cooling, 28, 96
solar heating, 26-62, 64, 72, 80, 95, 96, 106, 110-24; cells, 27; collectors, 26-7, 29-37, 39-59, 95, 105, construction of, 47-59, location of, 31, 41, 43-7, mounting of, 45-7; concentrators, 27-8; costs, 59-62; efficiency, 32, 61; stills, 28
solid fuel heating, 6, 17, 63-4, 67, 97
soundproofing, 7, 10, 20-1
storage problems, 37-44, 72, 78-9, 89, 93, 105
stovepipes, 6, 67, 70-1
sulphur dioxide, 64, 91
swimming pools, 29, 38, 60
sycamore wood, 66

tanks, 13, 14-17; hot-water, 5, 21-2, 37-43, 96, 105; underfloor, 38; underground, 29, 32
tax concessions, 27, 61
thermostats, 5, 23-4, 41, 69, 93
thermosyphon system, 34, 41
tides, 72
time switches, 23-4
turbines, 80-1, 84, 86-9; impulse, 87-9; reaction, 88

urea formaldehyde, 11
USA, 68, 78; — Forest Service, 64

ventilation, 7, 9, 10, 14, 17-18, 21, 68
Venturi, 78
Vermiculite, 13

Walker, Robert, 78
walls *see* insulation
Warmheart heating system, 97-9
waste disposal, 92
water capacity, 32, 33, 39, 59
water wheels, 80-1, 84-6; overshot, 86; undershot, 85-6; *see also individual headings*
waterpower, 80-9, 104, 107; measurement of, 81-5
weather, 12, 26, 60
weirs, 82, 83
'White Meter', 21, 23
windmills, 72, 73-8, 104, 105; horizontal drive, 73-5; verticle drive, 73-5; *see also individual headings*
windows, 7, 10, 17, 18, 19-21
windpower, 72-9, 106; speed, 72-4
wood, 50, 63-71; supply, 64-6; *see also individual headings*